Engineering Capstone Design Workbook

Volume 2

Build, Test, Redesign

Bahram Nassersharif, Ph.D.
Distinguished University Professor
Department of Mechanical, Industrial, and Systems Engineering
University of Rhode Island

Copyright © 2024 by Bahram Nassersharif.

All rights reserved. No part of this book may be reproduced or transmitted in any form or by any means, electronic or mechanical, including photocopying, recording, or any information storage and retrieval system, without permission in writing from the publisher.

The author and publisher have diligently applied their best efforts in the creation of this book, including the development, research, and evaluation of theories and programs to assess their effectiveness. However, the author and publisher provide no warranties, either expressed or implied, regarding the accuracy or applicability of the content, programs, or documentation within this book. Neither the author nor the publisher will be liable for any incidental or consequential damages arising from the provision, performance, or use of the materials included in this publication.

Table of Contents

15 BUILD A MORE COMPLETE PROTOTYPE 17
15.1 INTRODUCTION 17
- 15.1.1 Purpose of the Working Model 17
- 15.1.2 Approaches to Building the Working Model 17
- 15.1.3 Planning for the Working Model 18
- 15.1.4 Documentation and Reporting 19
- 15.1.5 Transition to Final Product 19

15.2 LOGISTICS OF BUILDING THE DESIGN 20
- 15.2.1 Resource Planning and Management 20
- 15.2.2 Developing the Build Plan 21
- 15.2.3 Execution of the Build Plan 22
- 15.2.4 Budget Management 22
- 15.2.5 Addressing Specific Project Needs 23
- 15.2.6 Final Considerations 23

15.3 RESOURCES NEEDED 24
- 15.3.1 Software Development Projects 24
- 15.3.2 Provision of Resources 25
- 15.3.3 Product Development Projects 25
- 15.3.4 Coordination and Scope 26
- 15.3.5 Common Resources for All Projects 26
- 15.3.6 Practical Time Management 26

15.4 ACTUAL VS. MODEL 27
- 15.4.1 Actual Implementation 27
- 15.4.2 Model Implementation 28
- 15.4.3 Decision-Making Criteria 29
- 15.4.4 Integration of Actual and Model Approaches 30

15.5 MATERIALS 31
- 15.5.1 Factors in Materials Selection 31
- 15.5.2 Mechanical Properties 31
- 15.5.3 Fabrication Requirements 32
- 15.5.4 Cost 32
- 15.5.5 Chemical Compatibility 33
- 15.5.6 Bill of Materials (BOM) 33
- 15.5.7 Materials Selection Charts 33

		15.5.8	Examples of Material Selection Charts:	34

- 15.5.8 Examples of Material Selection Charts: 34
- 15.5.9 Metals 34
- 15.5.10 Plastics 35
- 15.5.11 Ceramics 35
- 15.5.12 Composites 36
- 15.5.13 Coatings and Adhesives 36
- 15.5.14 Safety and Regulatory Considerations 37
- 15.5.15 Materials Selection Charts 37
- 15.6 MANUFACTURABILITY 42
  - 15.6.1 Key Parameters in Manufacturability 42
  - 15.6.2 Methods of Manufacturing 43
  - 15.6.3 Early Design Considerations 44
  - 15.6.4 Manufacturing Process Selection 45
  - 15.6.5 Creating a Project Plan for Manufacturing 48
  - 15.6.6 Design for Manufacturing and Assembly (DFMA) 49
  - 15.6.7 Mistake Proofing 52
  - 15.6.8 3D Printing 57
  - 15.6.9 Quality 63
- 15.7 PROCUREMENT 71
  - 15.7.1 University Procurement Policies 71
  - 15.7.2 Budget Management 71
  - 15.7.3 Internal Process for Procurement 71
  - 15.7.4 Steps in the Procurement Process 72
  - 15.7.5 Best Practices for Effective Procurement 73
  - 15.7.6 Example of a Procurement Workflow 74
- 15.8 ASSIGNMENTS 75

16 TEST ENGINEERING 81

- 16.1 TEST PLANNING 83
  - 16.1.1 Importance of Test Planning 83
  - 16.1.2 Components of a Test Plan 83
  - 16.1.3 Methodologies in Test Planning 84
  - 16.1.4 Best Practices in Test Planning 84
  - 16.1.5 Example of a Test Plan 85
- 16.2 TEST PLAN DEVELOPMENT 86
  - 16.2.1 Importance of Test Plan Development 86
  - 16.2.2 Components of a Test Plan 87
  - 16.2.3 Methodologies in Test Plan Development 88
  - 16.2.4 Best Practices in Test Plan Development 89
  - 16.2.5 Example of a Test Plan 89
- 16.3 TEST ENGINEERING MATRIX 92
  - 16.3.1 Importance of a Test Engineering Matrix 92
  - 16.3.2 Components of a Test Engineering Matrix 92
  - 16.3.3 Developing a Test Engineering Matrix 93
  - 16.3.4 Example of a Test Engineering Matrix 94
  - 16.3.5 Best Practices for Using a Test Engineering Matrix 98

		16.4	Test the Design and the Build	99
		16.4.1	Importance of Prototype Testing	99
		16.4.2	Methodologies in Prototype Testing	99
		16.4.3	Steps in Testing the Prototype	101
		16.4.4	Detailed Example of Prototype Testing	102
		16.4.5	Best Practices in Prototype Testing	105
	16.6		Assignments	107

17 REDESIGN 111

	17.1		Introduction to Redesign	111
		17.1.1	Iterative Nature of Design	111
		17.1.2	Continuous Improvement and Practical Constraints	112
		17.1.3	Capturing Requirements and Enhancements	112
		17.1.4	Revisiting Design Specifications	112
		17.1.5	Iterative Testing and Assessment	112
		17.1.6	Fundamental Design Flaws	113
		17.1.7	Quick Testing and Iterative Refinement	113
	17.2		Application of Test Results	113
		17.2.1	The Role of the Test Matrix	113
		17.2.2	Addressing Multiple Specifications	115
		17.2.3	Incorporating Sponsor and Customer Feedback	115
		17.2.4	Budget and Resource Considerations	116
		17.2.5	Documenting and Updating the Project Plan	116
		17.2.6	Real-World Implications and Continuous Improvement	116
	17.3		Adjustments to Design-Build	117
		17.3.1	Iterative Nature of Design Adjustments	117
		17.3.2	Dealing with Over-Constrained Systems	118
		17.3.3	Role of Stakeholders in Design Adjustments	118
		17.3.4	Practical Considerations for Design Adjustments	119
	17.4		Major Design Changes	120
		17.4.1	Reasons for Major Design Changes	120
		17.4.2	Process of Implementing Major Design Changes	121
		17.4.3	Implications of Major Design Changes	122
		17.4.4	Example Scenario	123
	17.5		Optimization	124
		17.5.1	Definition of Optimization	124
		17.5.2	Types of Optimization Problems	124
		17.5.3	Optimization Methods	125
		17.5.4	Optimization in Design Process	126
		17.5.5	Practical Application of Optimization	126
		17.5.6	Challenges in Optimization	127
		17.5.7	Optimization Software Tools	127
	17.6		Quality Engineering	128
		17.6.1	Garvin's Eight Dimensions of Quality	128
		17.6.2	Application of Quality Engineering in Design	130
		17.6.3	Tools and Techniques for Quality Engineering	131

17.7	COST EVALUATION		132
	17.7.1	Importance of Cost Evaluation	133
	17.7.2	Elements of Cost Evaluation	133
	17.7.3	Tools and Methods for Cost Evaluation	134
	17.7.4	Impact of Redesign on Cost Evaluation	135
	17.7.5	Example of Cost Evaluation	135
17.8	RETURN ON INVESTMENT		136
	17.8.1	Economic Analysis and Future Value of Money	136
	17.8.2	Practical Application	139
17.9	DESIGN OPTIMIZATION		140
	17.9.1	Purpose of Design Optimization	140
	17.9.2	Methods of Design Optimization	141
	17.9.3	Objective Function and Design Parameters	142
	17.9.4	Managing the Optimization Process	142
	17.9.5	Practical Example: Capstone Design Optimization	142
17.11	ASSIGNMENTS		144

18 OTHER CONSIDERATIONS ... 151

18.1	SAFETY		152
	18.1.1	Risk Assessment and Management	153
	18.1.2	Inherently Safe Design	154
	18.1.3	Redundancy and Fail-Safe Design	154
	18.1.4	Human Factors and Safety	155
	18.1.5	Safety Standards and Regulations	156
	18.1.6	Safety in the Product Lifecycle	156
	18.1.7	Emerging Technologies and Safety Challenges	156
18.2	ERGONOMICS		157
	18.2.1	Physical Ergonomics	158
	18.2.2	Cognitive Ergonomics	158
	18.2.3	Organizational Ergonomics	159
	18.2.4	Universal Design and Accessibility	160
	18.2.5	Evaluation and Testing in Ergonomic Design	160
	18.2.6	Emerging Trends in Ergonomics	161
18.3	HEALTH CONSIDERATIONS		162
	18.3.1	Physical Health Considerations	162
	18.3.2	Mental Health Considerations	163
	18.3.3	Health-Promoting Design	164
	18.3.4	Inclusive Health Design	165
	18.3.5	Health Monitoring and Feedback Systems	165
	18.3.6	Evaluation and Testing for Health Impacts	165
	18.3.7	Ethical Considerations in Health-Related Design	166
	18.3.8	Interdisciplinary Collaboration	166
18.4	ENVIRONMENTAL CONSIDERATIONS		167
	18.4.1	Life Cycle Assessment (LCA)	167
	18.4.2	Circular Economy Design	168
	18.4.3	Energy Efficiency and Renewable Energy Integration	168

- 18.4.4 Water Conservation and Management 169
- 18.4.5 Material Selection and Resource Efficiency 170
- 18.4.6 Biodiversity and Ecosystem Protection 170
- 18.4.7 Pollution Prevention and Control 171
- 18.4.8 Climate Change Mitigation and Adaptation 171
- 18.4.9 Waste Reduction and Management 172
- 18.4.10 Environmental Impact Assessment and Monitoring 172

18.5 Social Considerations 173
- 18.5.1 Understanding Social Impact 173
- 18.5.2 Participatory Design and Community Engagement 174
- 18.5.3 Cultural Sensitivity and Diversity 174
- 18.5.4 Accessibility and Universal Design 175
- 18.5.5 Social Equity and Justice 175
- 18.5.6 Technology and Social Change 176
- 18.5.7 Social Sustainability 176
- 18.5.8 Education and Capacity Building 176
- 18.5.9 Health and Well-being 177

18.6 Ethical Considerations 178
- 18.6.1 Fundamental Ethical Principles in Engineering 178
- 18.6.2 Safety and Risk 178
- 18.6.3 Environmental Ethics 179
- 18.6.4 Social Justice and Equity 180
- 18.6.5 Privacy and Data Ethics 180
- 18.6.6 Professional Integrity and Whistleblowing 181
- 18.6.7 Emerging Technologies and Future Ethics 181
- 18.6.8 Ethical Decision-Making Frameworks 182
- 18.6.9 Ethics Education and Professional Development 182

18.7 Political Considerations 183
- 18.7.1 Regulatory Compliance and Policy Frameworks 183
- 18.7.2 Public Infrastructure and Urban Planning 184
- 18.7.3 Technology Policy and Innovation 184
- 18.7.4 Geopolitical Considerations 185
- 18.7.5 Social Movements and Public Opinion 185
- 18.7.6 Strategies for Navigating Political Considerations 186

18.8 Sustainability 186
- 18.8.1 Defining Sustainability in Engineering Design 187
- 18.8.2 Key Principles of Sustainable Engineering Design 187
- 18.8.3 Strategies for Implementing Sustainability in Engineering Design 188
- 18.8.4 Challenges in Sustainable Engineering Design 188
- 18.8.5 Case Studies in Sustainable Engineering Design 189
- 18.8.6 The Future of Sustainability in Engineering Design 189

18.10 Assignments 191

19 Documentation of Capstone Design Projects 241

19.1 Introduction 241
- 19.1.1 Importance of Formal Documentation 241

	19.1.2 Engineering Design Documentation	242
	19.1.3 Guidelines for Effective Documentation	243
	19.1.4 Design Binder Implemented as a Shared Electronic Drive	243
	19.1.5 Electronic Files and Project Archive	246
19.2	VERBAL PRESENTATION WITH SLIDES	247
	19.2.1 Tips for Preparing an Excellent Presentation	247
	19.2.2 Guidelines for Capstone Design Presentations	249
19.3	PHOTOS	251
	19.3.1 Importance of Photographs in Engineering Design	251
	19.3.2 Practical Applications of Photographs in Capstone Design Projects	252
	19.3.3 Photo Example	252
19.4	VIDEOGRAPHY	254
	19.4.1 Planning and Goals for Project Videos	254
	19.4.2 Timing Studies	254
	19.4.3 Position, Velocity, and Acceleration Measurement	255
	19.4.4 Practical Considerations for Effective Videography	256
	19.4.5 Example Applications of Videography in Capstone Design Projects	256
19.5	POSTER	257
	19.5.1 Elements of a Design Poster	257
	19.5.2 Presentation Format	258
19.6	TECHNICAL INFORMATION SHEET	261
	19.6.1 Purpose and Importance of the Technical Information Sheet	261
	19.6.2 Key Components of a Technical Information Sheet	261
	19.6.3 Guidelines for Capstone Technical Information Sheets	262
	19.6.4 Style Guide for Capstone Technical Information Sheet	263
	19.6.5 Example Layout of a CTIS	263
19.7	FINAL DESIGN REPORT	267
	19.7.1 Guidelines for the Final Design Report	267
	19.7.2 Professional Formatting Guidelines	270
19.9	ASSIGNMENTS	274

BIBLIOGRAPHY 283

INDEX 289

List of Figures

FIGURE 15-1. RELATIONSHIP OF MATERIALS TO PROCESS, FORM, AND FUNCTION. 32

FIGURE 15-2. MATERIAL SELECTION CHART – YOUNG'S MODULUS VERSUS DENSITY. 38

FIGURE 15-3. MATERIAL SELECTION CHART – YOUNG'S MODULUS VERSUS COST. 39

FIGURE 15-4. MATERIAL SELECTION CHART – STRENGTH VERSUS TOUGHNESS. 39

FIGURE 15-5. MATERIAL SELECTION CHART – STRENGTH VERSUS COST. 40

FIGURE 15-6. MATERIAL SELECTION CHART – STRENGTH VERSUS ELONGATION. 40

FIGURE 15-7. MATERIAL SELECTION CHART – STRENGTH VERSUS MAXIMUM TEMPERATURE. 41

FIGURE 15-8. MATERIAL SELECTION CHART – STRENGTH VERSUS TOUGHNESS. 41

FIGURE 15-9. POKA-YOKE EXAMPLE USB-A CONNECTOR. 54

FIGURE 15-10. POKA-YOKE EXAMPLE USB-MICRO CONNECTOR. 54

FIGURE 15-11. POKA-YOKE EXAMPLE USB-C CONNECTOR. 55

FIGURE 15-12. POKA-YOKE EXAMPLE USB-C CONNECTOR. 55

FIGURE 15-13. 3D PRINTING PROCESS. 61

FIGURE 19-1. EXAMPLE PHOTOGRAPH SHOWING MAGNETIC SEAL DEMAGNETIZER. 253

FIGURE 19-2. EXAMPLE POSTER. 259

FIGURE 19-3. EXAMPLE CTIS FRONT SIDE. 265

FIGURE 19-4. EXAMPLE CTIS BACK SIDE. 266

List of Tables

TABLE 15-1. EXAMPLE BILL OF MATERIALS TABLE. 69

TABLE 17-1. EXAMPLE OF A TEST MATRIX FOR A CAPSTONE PROJECT. 114

List of Assignments

ASSIGNMENT 18-1. ENGINEERING ETHICS DIAGNOSTIC QUIZ 196

ASSIGNMENT 19-1: CREATE A POSTER FOR YOUR DESIGN PROJECT 274

ASSIGNMENT 19-2: CREATE A TECHNICAL INFORMATION SHEET FOR YOUR DESIGN PROJECT 278

Preface

The journey through engineering design is one of iteration, innovation, and continuous refinement. While Volume 1 of the "Engineering Capstone Design Workbook" focused on the foundational aspects—problem definition, concept creation, and initial validation—Volume 2 extends this path into the realms of detailed design, prototype development, and iterative improvements. This volume is designed to accompany students through the later stages of their capstone projects, where ideas evolve into tangible products that meet or exceed defined specifications.

Volume 2 emphasizes the creation of complete working prototypes and introduces advanced methodologies for engineering testing and validation. Students will develop test engineering plans and execute them to verify that their solutions align with both the original design specifications and the customer's needs. These tests often reveal the necessity for design adjustments, making iterative redesigns a critical component of the process. Through these refinements, students will learn to optimize their prototypes, ensuring that each iteration brings the design closer to a viable, customer-ready solution.

The iterative nature of engineering design lies at the heart of this volume. It mirrors real-world engineering practices where testing, feedback, and corrections are repeated until an optimal version of the product emerges. This approach fosters a mindset of perseverance and adaptability—two essential qualities for successful engineers.

Key topics covered in Volume 2 include the integration of mechanical, electrical, and software systems into a unified prototype, advanced testing techniques, and methods for managing project timelines and budgets during iterative redesigns. The volume also emphasizes collaboration and communication across multidisciplinary teams, preparing students to navigate the complexities of professional engineering environments.

By the conclusion of Volume 2, students will not only have a working prototype but will also have developed the practical skills and confidence needed to deliver high-quality engineering solutions. This volume completes the capstone experience, transforming concepts into products that address real-world problems and meet customer expectations.

Bahram Nassersharif, Ph.D.

October 2024

15 Build a More Complete Prototype

15.1 Introduction

The transition from proof of concept (POC) to a more advanced working model is a critical phase in the design process. The POC design is typically substantiated with a prototype that validates key aspects of the overall design. The next step is to refine and develop the design further, creating a more functional and usable model. This working model may be a physical build in the case of a product or a process model implemented in simulation software such as ProModel.

15.1.1 Purpose of the Working Model

The primary purpose of developing a working model is to advance the design concept by addressing the limitations and lessons learned from the POC prototype. This stage involves:

1. **Refinement of Design**: Enhancing the design based on feedback and testing results from the POC prototype.
2. **Functional Usability**: Creating a model that is not only functional but also practical for end-users.
3. **Validation and Verification**: Conducting thorough testing to ensure the model meets all design specifications and requirements.

15.1.2 Approaches to Building the Working Model

The development of the working model can follow two primary approaches, depending on the insights gained from the POC prototype:

1. **Evolution of the POC Prototype**

- **Natural Progression**: This approach involves iteratively improving the existing POC prototype. The design team builds upon the initial prototype, refining its features and functionality to create a more mature and robust model.
- **Incremental Enhancements**: Each iteration focuses on enhancing specific aspects of the design, such as material selection, component integration, or performance optimization.

2. **Fresh Start**
   - **Reevaluation and Redesign**: In some cases, the insights gained from the POC prototype indicate the need for a significant redesign. The design team may start afresh, using the lessons learned to develop a new and improved working model.
   - **New Concepts and Innovations**: This approach allows the team to incorporate new ideas and innovations that were not feasible or considered during the POC stage.

15.1.3 Planning for the Working Model

Developing a working model requires careful planning and execution. The following steps outline the key elements of this planning process:

1. **Define Objectives and Requirements**
   - Clearly articulate the goals for the working model, including specific functionalities, performance metrics, and user requirements.
   - Ensure that these objectives align with the overall project goals and sponsor expectations.
2. **Develop a Project Plan**
   - **Tasks and Milestones**: Break down the development process into manageable tasks and define key milestones. Each task should have a clear objective, timeline, and responsible team member.
   - **Resource Allocation**: Identify the resources needed for each task, including materials, equipment, and personnel. Allocate resources efficiently to ensure timely completion of each task.
   - **Risk Management**: Assess potential risks and challenges that may arise during the development process. Develop mitigation strategies to address these risks proactively.
3. **Design and Development**
   - **Detailed Design**: Create detailed design specifications and drawings for the working model. Ensure that all components and subsystems are thoroughly defined and documented.
   - **Prototyping and Fabrication**: Depending on the nature of the

working model, proceed with prototyping or fabrication. This may involve 3D printing, machining, or assembly of components.
    - **Integration and Testing**: Integrate all components and subsystems into the working model. Conduct rigorous testing to validate the design against the defined objectives and requirements.
4. **Iterative Improvement**
    - **Feedback and Evaluation**: Gather feedback from testing and end-users. Evaluate the performance of the working model and identify areas for improvement.
    - **Continuous Refinement**: Implement iterative improvements based on feedback and testing results. This may involve multiple cycles of refinement and testing to achieve the desired level of functionality and usability.

15.1.4 Documentation and Reporting

Throughout the development of the working model, it is essential to maintain thorough documentation and reporting:

1. **Design Documentation**
    - **Design Specifications**: Update design specifications to reflect any changes or enhancements made during the development process.
    - **Drawings and Schematics**: Ensure that all design drawings and schematics are current and accurately reflect the working model.
2. **Testing Reports**
    - **Test Plans and Procedures**: Document the testing plans and procedures used to validate the working model.
    - **Test Results and Analysis**: Record the results of all tests conducted, along with any analysis and conclusions drawn from these results.
3. **Progress Reports**
    - **Project Updates**: Provide regular updates on the progress of the working model development to stakeholders, including sponsors, mentors, and team members.
    - **Milestone Achievements**: Highlight key milestones achieved and any significant findings or accomplishments.

15.1.5 Transition to Final Product

Once the working model is validated and meets all design specifications, the next step is to transition towards the final product:

1. **Production Planning**

- Develop a detailed plan for scaling up production, including manufacturing processes, quality control, and supply chain management.
2. **User Training and Support**
    - Prepare user manuals, training materials, and support documentation to ensure that end-users can effectively use and maintain the final product.
3. **Final Validation and Certification**
    - Conduct final validation and certification tests to ensure compliance with regulatory standards and industry requirements.

By following these guidelines, the transition from a proof of concept to a fully functional working model can be effectively managed, ensuring a robust and reliable final product. This phase is critical in bridging the gap between initial design concepts and practical, real-world applications.

15.2 Logistics of Building the Design

Building the design concept beyond the prototype requires meticulous planning and allocation of resources. This section delves into the logistical aspects of advancing a design from proof of concept to a functional working model, emphasizing the importance of a comprehensive design plan, resource management, and stakeholder coordination.

15.2.1 Resource Planning and Management

The successful development of a working model hinges on the effective management of several key resources:

15.2.1.1 People

- **Team Members**: Ensure that all team members have clearly defined roles and responsibilities. This includes design, engineering, procurement, assembly, testing, and documentation.
- **Stakeholders**: Coordinate with sponsors, mentors, and capstone professors for guidance, feedback, and approvals.

15.2.1.2 Time

- **Timeline**: Develop a detailed project timeline that outlines each phase of the build process, from initial planning to final testing. Include all critical milestones and deadlines.
- **Scheduling**: Allocate sufficient time for each task, ensuring that there is buffer time for unexpected delays or issues.

15.2 Logistics of Building the Design

15.2.1.3 Materials

- **Bill of Materials (BOM)**: Create a comprehensive BOM listing all materials required for the build. Include specifications, quantities, suppliers, and estimated costs.
- **Procurement**: Plan the procurement process, including submitting purchase requests, receiving approvals, ordering, shipping, and receiving materials.

15.2.1.4 Software and Instruments

- **Software Tools**: Identify and secure access to necessary software tools for modeling, simulation, and design validation.
- **Instruments**: Ensure availability of any specialized instruments required for testing and measurements.

15.2.1.5 Facilities

- **Build Space**: Secure appropriate build space based on the size and nature of the project. This could range from a computer lab for software projects to a workshop for physical prototypes.
- **Shared Resources**: Schedule access to shared facilities such as machine shops, electronics labs, 3D printing facilities, welding stations, and testing equipment well in advance to avoid conflicts.

15.2.2 Developing the Build Plan

A well-developed build plan is essential for the successful execution of the project:

15.2.2.1 Scope Definition

- Define the scope of the build to align with project requirements and available resources. Ensure the plan allows for iterative improvements and does not exhaust all resources on the initial build.

15.2.2.2 Stakeholder Review

- Present the proposed build plan to sponsors and the capstone professor for review. Discuss anticipated challenges and resolve any issues to the satisfaction of all stakeholders.

15.2.2.3 Phased Approach

- Adopt a phased approach to the build process, allowing for incremental improvements and testing. This approach helps manage risks and ensures that any issues can be addressed early.

15.2.3 Execution of the Build Plan

The timeline for the build phase, testing, and redesign is crucial, with practical time allocations as follows:
- **Build Phase**: 6 weeks
- **Testing**: 2 weeks
- **Redesign and Wrapping Up**: 5 weeks

15.2.3.1 Material Procurement

- o **Cost Efficiency**: Shop around for the best possible prices for materials. Consider bulk purchasing and negotiate with suppliers for discounts.
- o **Tracking**: Maintain a procurement log to track the status of purchase requests, approvals, orders, shipments, and receipts.

15.2.3.2 Scheduling and Coordination

- o **Lab and Resource Scheduling**: Book lab space, equipment, and other resources as early as possible. Create a detailed schedule to ensure optimal use of shared resources.
- o **Team Coordination**: Hold regular meetings to coordinate tasks, monitor progress, and address any issues promptly.

15.2.3.3 Assembly and Fabrication

- o **Workspace Setup**: Ensure the workspace is organized and equipped with all necessary tools and materials before starting the assembly.
- o **Step-by-Step Assembly**: Follow the build plan meticulously, documenting each step and making adjustments as needed based on real-time observations and feedback.

15.2.4 Budget Management

15.2.4.1 Initial Budgeting

- o Develop a comprehensive budget that covers all aspects of the build, including materials, labor, equipment, and contingency funds for unforeseen expenses.

15.2.4.2 Ongoing Monitoring

- o Regularly review and update the budget to reflect actual expenditures. Monitor costs closely to ensure that the project stays within budget.

15.2 Logistics of Building the Design

15.2.4.3 Future Considerations
- Plan for additional expenses related to testing, redesign, and final validation. Allocate funds to cover potential improvements and iterations based on testing results.

15.2.5 Addressing Specific Project Needs

15.2.5.1 Software Development Projects
- **Digital Resources**: Ensure adequate computer and cloud storage resources. Secure licenses for any necessary software tools.
- **Physical Space**: Arrange for a dedicated space such as a computer lab or meeting room for team collaboration and development activities.

15.2.5.2 Process Design Projects
- **Data Collection**: Coordinate with sponsors to access process data or experiment with design concepts. If direct access is not feasible, develop software models or create mock processes for testing.
- **Simulation Tools**: Use modeling and simulation software like ProModel to refine and validate process designs.

15.2.5.3 Product Development Projects
- **Physical Space**: Determine the space requirements based on the size of the working model. Large projects may require substantial floor space and specialized equipment.
- **Technical Resources**: Arrange access to machine shops, electronics labs, 3D printers, welding stations, and testing facilities. Schedule these resources in advance to prevent delays.

15.2.6 Final Considerations

15.2.6.1 Iterative Improvement
- Plan for multiple build iterations, allowing for continuous improvement based on testing and feedback.

15.2.6.2 Documentation
- Maintain detailed documentation throughout the build process, including design changes, testing results, and any issues encountered. This documentation will be crucial for future iterations and final reporting.

15.2.6.3   Stakeholder Engagement
- Keep stakeholders informed and engaged throughout the build process. Regular updates and feedback sessions help ensure alignment with project goals and sponsor expectations.

By meticulously planning and managing the logistics of building the design within the given timeline, the design team can effectively transition from a proof of concept to a fully functional working model. This phase is crucial for validating the design, identifying areas for improvement, and ensuring that the final product meets all requirements and specifications.

15.3 Resources Needed

Once a plan for the build of the design has been created and reviewed by stakeholders, the design team must identify all resources needed to complete the task. Resources needed vary depending on whether the team plans to construct software, a process, or a product prototype. Below are the specific resources required for each type of project:

15.3.1   Software Development Projects

For a software development project, the team may need the following resources:

1. **Computer Systems**
    - High-performance computers for development and testing.
    - Access to servers for deploying and running software in a networked environment.
2. **Operating System**
    - Selection of an operating system (Windows, Mac OS, Linux, Unix, etc.) that aligns with the project requirements.
3. **Software Development Environment**
    - Integrated Development Environment (IDE) such as Visual Studio, Eclipse, IntelliJ IDEA, or PyCharm.
    - Compilers, interpreters, and necessary libraries.
    - Version control systems like Git for source code management.
4. **User Interface System**
    - Tools and frameworks for building user interfaces, which could be open-source or vendor-specific (e.g., Windows, Mac OS, iOS, Android, Unix, Chrome OS).

15.3.2 Provision of Resources

These resources are typically provided by the school or the project sponsor. Students may also use their personal hardware and software for development.

1. **Process Design Projects**

For process design projects involving new development or modifications to existing systems, the following resources are required:

2. **Access to Current Process**
   - Permission to review and document the existing process to collect data and create a dynamic model.
3. **Transportation**
   - Arrangements for travel to and from the facility where the process is located, if applicable.
4. **Personnel Access**
   - Access to individuals involved in the current process for interviews and information gathering to understand the process details and expected changes.
5. **Simulation Software**
   - Software tools for process simulation such as ProModel, MATLAB, Visual Components, MapleSIM, etc.
6. **Video Capture Hardware**
   - Cameras or smartphones for recording videos of the current process for detailed analysis.
7. **Computing Resources**
   - High-performance computers capable of running simulation and modeling software efficiently.

15.3.3 Product Development Projects

For product development projects, the decision to scale the working model is crucial. The team must decide whether to create a full-scale model subjected to real-world conditions or a scaled-down version. This decision is influenced by budget, time constraints, and the specific requirements of the project. Typical resources needed include:

1. **Materials**
   - Raw materials in basic shapes such as rectangular bars, sheets, rods, tubes, and squares. The design must utilize these elemental shapes for practicality and availability.
2. **Tools and Equipment**
   - Access to tools and machinery such as saws, drills, sanders, milling

machines, lathes, welders, and soldering equipment.
3. **Fasteners and Connectors**
    - Screws, nuts, bolts, clamps, and pins for assembling components.
4. **Adhesives and Sealants**
    - Appropriate adhesives and sealants for joining materials and ensuring environmental durability.
5. **3D Printing**
    - 3D printers and materials such as PLA, TPU, PC, PLA+, and ABS for creating complex parts and prototypes.

15.3.4 Coordination and Scope

- The scope of the working model should be coordinated with the sponsor and other stakeholders. Decisions regarding full-size versus scaled-down models depend on budget constraints and time availability.

15.3.5 Common Resources for All Projects

Regardless of the project type, certain resources are universally required:

1. **Lab Space**
    - Designated lab space for group work, assembly, and testing of the project.
2. **Meeting Space**
    - A meeting area for regular interactions with sponsors, mentors, and stakeholders to discuss progress and gather feedback.
3. **Project Management Tools**
    - Software for project management and collaboration, such as Trello, Asana, or Microsoft Project.
4. **Documentation Tools**
    - Tools for documentation and reporting, such as Microsoft Office Suite or Google Workspace, to maintain records of the design process, testing results, and iterations.

15.3.6 Practical Time Management

Given the constrained timeline typical of capstone projects, careful time management is essential:

1. **Build Phase (6 weeks)**
    - Allocate the first six weeks to the assembly and construction of the working model. Ensure that material procurement, lab scheduling, and team coordination are completed efficiently to maximize build time.

2. **Testing Phase (2 weeks)**
   - Dedicate two weeks to rigorous testing of the working model. This phase includes performance validation, stress testing, and compliance checks with design specifications.
3. **Redesign and Finalization (5 weeks)**
   - Use the final five weeks for redesign based on testing feedback and wrapping up the project. This includes final adjustments, documentation, and preparation for the final presentation and report submission.

Properly identifying and managing resources is critical for the success of capstone design projects. Whether the project involves software development, process design, or product development, the availability and effective utilization of resources such as materials, tools, software, and facilities significantly impact the project's outcome. By carefully planning and coordinating with stakeholders, design teams can navigate the logistical challenges and successfully transition from proof of concept to a fully functional working model.

15.4 Actual vs. Model

Capstone projects often involve complex real-world problems with numerous interdependencies and interfacing systems. The scope of these projects can encompass either the development of an actual product or process or a more simplified, isolated, and smaller-scale implementation that aligns with the budget and schedule constraints of the capstone program. Understanding the distinctions between working with an actual full-scale model and a scaled-down or simulated model is crucial for the success of the project.

15.4.1 Actual Implementation

Building the actual product or process involves developing a full-scale version that operates under real-world conditions. This approach offers several advantages and challenges:

15.4.1.1 Advantages

1. **Realistic Testing and Validation**
   - Provides an accurate representation of the final product's performance, allowing for thorough testing and validation under real-world conditions.
   - Enables the identification and resolution of issues related to actual usage, such as material behavior, environmental impacts,

and user interactions.
2. **Stakeholder Confidence**
    - Demonstrates to stakeholders, including sponsors and potential users, that the design can function as intended in a practical setting.
    - Increases the credibility and perceived reliability of the design, which can be crucial for securing future support or funding.
3. **Regulatory and Compliance Testing**
    - Facilitates comprehensive testing for compliance with industry standards and regulatory requirements, ensuring the design meets all necessary certifications.

15.4.1.2 Challenges

1. **Resource Intensive**
    - Requires significant resources, including materials, time, and budget. The costs associated with building a full-scale model can be substantial.
    - Demands access to specialized tools, equipment, and facilities, which may be limited or shared with other teams.
2. **Complexity and Risk**
    - Involves managing a higher level of complexity due to the interdependencies and interactions with other systems.
    - Poses greater risks related to potential failures or unforeseen issues that could impact the project timeline and budget.
3. **Time Constraints**
    - The fixed period of the capstone design course may not be sufficient to develop, test, and refine a full-scale model, necessitating efficient time management and contingency planning.

15.4.2 Model Implementation

Creating a simplified, isolated, or smaller-scale model offers an alternative approach that can be more feasible within the constraints of a capstone project. This approach also has its own set of advantages and challenges:

15.4.2.1 Advantages

1. **Resource Efficiency**
    - Requires fewer materials and less budget, making it more cost-effective.
    - Reduces the need for specialized tools and facilities, allowing

for greater flexibility and ease of access.
2. **Focused Testing**
    - Enables targeted testing of specific design aspects or subsystems, facilitating a more manageable and controlled evaluation process.
    - Allows for rapid prototyping and iteration, which can be beneficial for refining design concepts.
3. **Scalability and Adaptability**
    - Provides a foundation for scaling up the design in the future. Insights gained from the model can guide the development of a full-scale version.
    - Offers the flexibility to adapt and modify the design based on testing feedback without significant resource expenditure.

15.4.2.2 Challenges

1. **Limited Realism**
    - May not fully capture the complexities and interdependencies of the actual product or process, potentially overlooking critical issues that could arise in real-world applications.
    - Validation and testing under scaled-down conditions may not be entirely indicative of full-scale performance.
2. **Stakeholder Perception**
    - Sponsors and stakeholders may perceive the model as less convincing or comprehensive compared to a full-scale implementation.
    - Demonstrating the practical viability of the design through a model may require additional explanation and justification.
3. **Iterative Development**
    - The process of scaling up from a model to a full-scale version involves additional iterations and potential redesigns, which can extend the overall project timeline.

15.4.3 Decision-Making Criteria

Deciding between actual implementation and model implementation involves several key considerations:

1. **Project Scope and Objectives**
    - Define the primary goals of the project and determine whether a full-scale or scaled-down approach aligns better with these objectives.

2. **Budget and Resources**
   o Assess the available budget and resources, including materials, equipment, and personnel. Consider the feasibility of each approach within these constraints.
3. **Time Constraints**
   o Evaluate the project timeline and the fixed period of the capstone course. Determine whether the team can realistically complete a full-scale implementation within the allotted time.
4. **Stakeholder Expectations**
   o Understand the expectations and requirements of sponsors and other stakeholders. Ensure that the chosen approach meets their needs and provides the necessary level of confidence in the design.
5. **Risk Management**
   o Identify potential risks associated with each approach and develop mitigation strategies. Consider the impact of these risks on the project's success and the ability to achieve the desired outcomes.
6. **Future Plans**
   o Consider the long-term goals for the design. If the project aims to develop a commercially viable product, starting with a scalable model may be more practical, allowing for future iterations and refinements.

15.4.4 Integration of Actual and Model Approaches

In many cases, integrating aspects of both actual and model implementations can provide a balanced approach:

1. **Hybrid Models**
   o Develop a hybrid model that combines full-scale components with scaled-down or simulated elements. This approach allows for focused testing while managing resource constraints.
2. **Phased Development**
   o Implement a phased development strategy, starting with a model to validate core concepts and gradually scaling up to a full-scale version as resources and time permit.
3. **Simulation and Prototyping**
   o Use simulation tools and rapid prototyping techniques to bridge the gap between model and actual implementation. Simulations can provide valuable insights into system behavior, while

prototypes can validate specific design aspects.

By carefully considering the advantages and challenges of actual versus model implementation and adopting a strategic approach, capstone design teams can effectively navigate the complexities of their projects. This ensures that the final design is robust, reliable, and meets the project's goals and stakeholder expectations.

15.5 Materials

The choice of materials plays a fundamental role in the product or device design process. Many facets must be considered, such as safety, strength, service requirements, fabrication requirements, cost, and chemical compatibility. By selecting sustainable and safe materials, designers can ensure the safety of humans and the environment, while also meeting performance requirements during the product's life and enabling recycling and reuse after its useful life. This section delves into the various considerations and types of materials commonly used in building designs, including metals, plastics, ceramics, and composites.

15.5.1 Factors in Materials Selection

15.5.1.1 Safety and Sustainability

- **Human Safety**: Materials must be non-toxic, fire-resistant, and safe to handle and use.
- **Environmental Safety**: The impact of materials on the environment should be minimal. Considerations include the carbon footprint, potential for recycling, and biodegradability.

15.5.2 Mechanical Properties

- **Strength**: The ability of a material to withstand an applied force without failure or plastic deformation.
- **Stiffness**: The rigidity of a material and its ability to resist deformation under load.
- **Toughness**: The ability to absorb energy and plastically deform without fracturing.
- **Hardness**: Resistance to localized plastic deformation or indentation.
  Service Requirements
- **Temperature Resistance**: Ability to maintain performance at high or low temperatures.
- **Corrosion Resistance**: Ability to withstand chemical or electrochemical reactions

with the environment.
- **Wear Resistance**: Ability to resist abrasion, erosion, and other forms of wear.

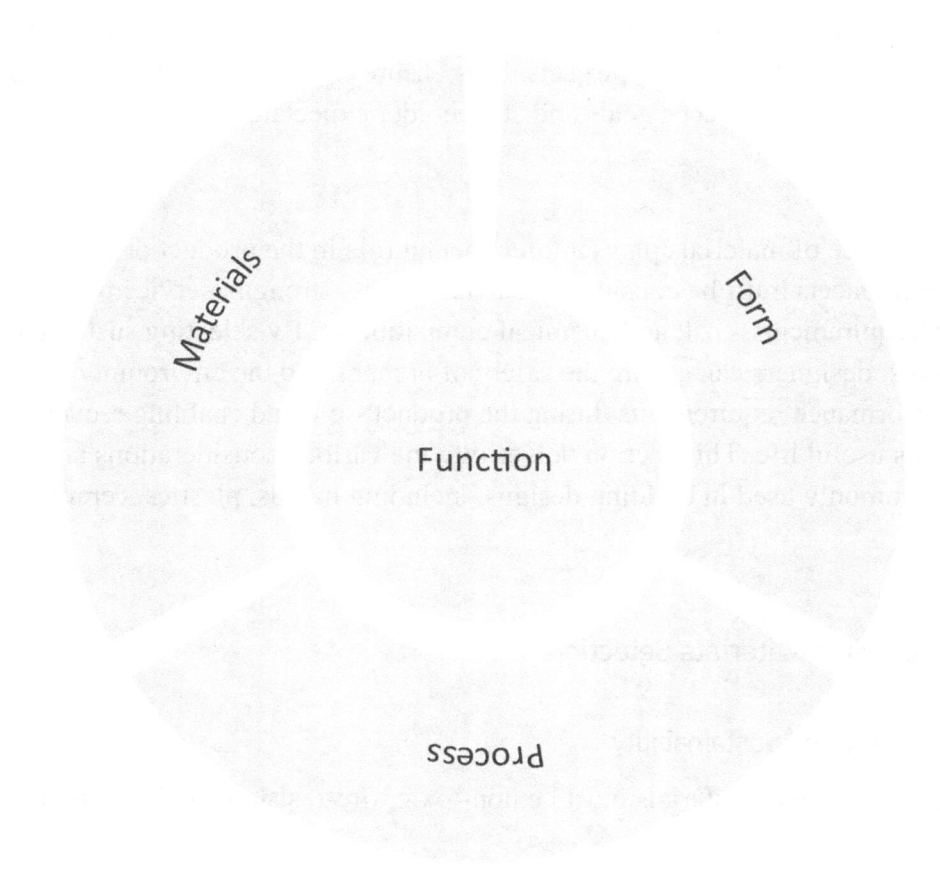

Figure 15-1. Relationship of materials to process, form, and function.

15.5.3 Fabrication Requirements
- **Machinability**: Ease with which a material can be cut, shaped, and finished.
- **Weldability**: Ability to be welded without defects.
- **Formability**: Ease with which a material can be formed into desired shapes.

15.5.4 Cost
- **Material Cost**: Direct cost of raw materials.
- **Processing Cost**: Costs associated with manufacturing processes, including labor and energy.
- **Lifecycle Cost**: Total cost of ownership, including maintenance, operation, and disposal costs.

15.5.5 Chemical Compatibility

- **Reactivity**: Interaction with other materials or environmental agents.
- **Solubility**: Ability to dissolve in various solvents.
- **Stability**: Resistance to degradation over time under service conditions.

15.5.6 Bill of Materials (BOM)

Product designers create a Bill of Materials (BOM) to outline all materials required for the design and build plans for the working model. The BOM includes detailed listings of all materials used in the design, ensuring that each part and assembly has an associated material specification. Typical materials include:

- **Wood and Wood Products**: Pine, balsa, oak, MDF.
- **Mesh Fabrics**: Kevlar, carbon fiber mesh.
- **Metals and Alloys**: Aluminum alloys, mild steel, alloy steels, stainless steels, cast iron, copper, brasses, nickel alloys, titanium alloys, magnesium alloys, zinc alloys, lead.
- **Polymers**: Nylon, PET, HDPE, PVC, polypropylene, POM, acetal, ABS, PLA, polyethylene, polystyrene, polycarbonate, porcelain.
- **Ceramics**: Alumina, brick, concrete, silicon, silicon carbide, diamond, titanium carbide, zirconia.
- **Composites**: Carbon fiber epoxy, fiberglass epoxy, mixed carbon fiber Kevlar composite.
- **Coatings and Paints**: Epoxy, paints.
- **Adhesives**: Glues, silicone, waterproof adhesives.

15.5.7 Materials Selection Charts

Material selection charts are a quick and powerful way to identify materials for design projects. These charts represent material properties graphically, typically using log-log scales to cover the extensive range of values. Common material property comparisons include:

- **Young's Modulus vs. Density**
- **Young's Modulus vs. Cost**
- **Material Strength vs. Density**
- **Material Strength vs. Toughness**
- **Material Strength vs. Strain**
- **Material Strength vs. Cost**
- **Material Strength vs. Temperature**
- **Specific Stiffness vs. Specific Strength**
- **Electrical Resistivity vs. Cost**

15.5.8 Examples of Material Selection Charts:

1. **Young's Modulus vs. Density**: This chart helps identify materials with high stiffness and low density, crucial for applications where weight is a primary consideration, such as in aerospace.
2. **Young's Modulus vs. Cost**: Useful for balancing performance with budget constraints, this chart can guide decisions on selecting cost-effective materials without compromising on stiffness.
3. **Material Strength vs. Toughness**: This chart differentiates materials based on their ability to resist deformation and absorb energy, helping select materials for applications requiring high impact resistance.
4. **Material Strength vs. Temperature**: Essential for high-temperature applications, this chart assists in choosing materials that maintain their strength at elevated temperatures.

15.5.9 Metals

Metals are often chosen for their strength, durability, and versatility in manufacturing. They are typically solid at room temperature, reflective, ductile, and good conductors of heat and electricity. Metals have high density, melting points, and boiling points, making them suitable for a wide range of applications, including high-temperature environments.

15.5.9.1 Common Metals and Their Properties:

1. **Steel**
    - **Hot-Rolled Steel**: Cost-effective, used in construction.
    - **Cold-Finished Steel**: Higher precision and strength, used in machining.
    - **Tool Steel**: High hardness and wear resistance, used for tools and dies.
2. **Aluminum**
    - **Aluminum Alloys (e.g., 6061, 6063)**: Lightweight, corrosion-resistant, used in aerospace and automotive applications.
3. **Stainless Steel**
    - **Alloys (e.g., 304, 316)**: Corrosion-resistant, high strength, used in medical devices and kitchenware.
4. **Copper**
    - Excellent thermal and electrical conductivity, used in electrical wiring and plumbing.
5. **Brass**
    - Alloy of copper and zinc, known for its machinability and corrosion resistance, used in fittings and valves.

15.5.10 Plastics

Plastics are versatile materials widely used in engineering due to their ease of fabrication, flexibility, and cost-effectiveness. They are available in various forms, including sheets, rods, tubes, and profiles, and can be easily cut, drilled, glued, and welded.

15.5.10.1 Common Plastics and Their Applications:

1. **ABS (Acrylonitrile Butadiene Styrene)**
   - Impact-resistant, used in automotive parts, toys, and consumer electronics.
2. **Nylon (Polyamide)**
   - High strength and flexibility, used in gears, bearings, and textiles.
3. **Polycarbonate**
   - High impact resistance and transparency, used in safety glasses, lenses, and electronic components.
4. **PVC (Polyvinyl Chloride)**
   - Durable and chemical-resistant, used in pipes, fittings, and flooring.
5. **PEEK (Polyether Ether Ketone)**
   - High-temperature resistance and mechanical strength, used in aerospace and medical implants.
6. **PLA (Polylactic Acid)**
   - Biodegradable and used in 3D printing and disposable packaging.

15.5.11 Ceramics

Ceramics are non-metallic, inorganic materials known for their hardness, high melting points, and resistance to wear and corrosion. They are often used in applications requiring high-temperature resistance and mechanical strength.

15.5.11.1 Common Ceramics and Their Applications:

1. **Alumina (Aluminum Oxide)**
   - High hardness and wear resistance, used in cutting tools and abrasives.
2. **Silicon Carbide**
   - High thermal conductivity and strength, used in semiconductor devices and heating elements.
3. **Zirconia (Zirconium Dioxide)**
   - High fracture toughness and thermal insulation, used in dental implants and thermal barriers.
4. **Titanium Carbide**
   - High hardness and wear resistance, used in cutting tools and coatings.

15.5.12 Composites

Composites combine two or more materials to achieve properties that are superior to those of the individual components. They are known for their high strength-to-weight ratio and are used in various high-performance applications.

15.5.12.1 Common Composites and Their Applications:

1. **Carbon Fiber Epoxy**
    - High strength and stiffness, used in aerospace, automotive, and sporting goods.
2. **Fiberglass Epoxy**
    - Durable and lightweight, used in boat hulls, pipes, and construction.
3. **Kevlar Composites**
    - High tensile strength and impact resistance, used in body armor and aerospace applications.

15.5.13 Coatings and Adhesives

Coatings and adhesives play a crucial role in protecting materials and assembling components. They enhance the durability, appearance, and performance of the final product.

15.5.13.1 Common Coatings:

1. **Epoxy Coatings**
    - Provide excellent chemical and corrosion resistance, used in flooring and protective coatings.
2. **Powder Coatings**
    - Durable and environmentally friendly, used in automotive parts and appliances.
3. **Paints**
    - Protect and decorate surfaces, available in various formulations for different substrates and environments.

15.5.13.2 Common Adhesives:

1. **Epoxy Adhesives**
    - High strength and chemical resistance, used in structural bonding and repairs.
2. **Silicone Adhesives**
    - Flexible and weather-resistant, used in sealing and gasketing.
3. **Cyanoacrylate (Super Glue)**
    - Quick-bonding and versatile, used in general-purpose bonding.

4. **Polyurethane Adhesives**
    - Strong and flexible, used in construction and automotive applications.

15.5.14 Safety and Regulatory Considerations

15.5.14.1 Material Safety Data Sheets (MSDS)

- MSDSs provide essential information on the safe handling, storage, and disposal of materials. They include data on chemical composition, health hazards, first aid measures, and fire-fighting instructions.

15.5.14.2 Regulatory Compliance

- Ensure all materials comply with relevant safety and environmental regulations, such as OSHA, REACH, and RoHS.

15.5.14.3 Fire Safety

- Maintain an inventory of materials stored in the lab and their corresponding MSDSs. Firefighters need this information to use appropriate fire retardants.

By carefully selecting materials based on the considerations outlined above, design teams can create safe, sustainable, and high-performing products that meet the needs of users and stakeholders.

15.5.15 Materials Selection Charts

The following material selection charts provide a visual representation of various material properties, helping in the decision-making process for selecting appropriate materials for design projects:

Figure 15-2. Material Selection Chart – Young's Modulus versus Density. This chart illustrates the relationship between Young's Modulus (a measure of stiffness) and density for different materials. It helps identify materials that offer high stiffness while maintaining low weight, which is crucial for applications like aerospace and automotive industries.

Figure 15-3. Material Selection Chart – Young's Modulus versus Cost. This chart compares the stiffness of materials with their cost, aiding in the selection of cost-effective materials without compromising on stiffness. It is useful for balancing performance and budget constraints in engineering projects.

Figure 15-4. Material Selection Chart – Strength versus Toughness. This chart shows the trade-off between material strength and toughness. It helps in selecting materials that can withstand high stress while absorbing energy without fracturing, making it ideal for impact-resistant applications.

Figure 15-5. Material Selection Chart – Strength versus Cost. This chart provides a comparison of material strength against cost, assisting in identifying materials that offer high strength at a reasonable price. It is crucial for optimizing the cost-performance ratio in design projects.

Figure 15-6. Material Selection Chart – Strength versus Elongation. This chart demonstrates the relationship between material strength and elongation, helping in the selection of materials that need to balance tensile strength with flexibility. It is important for applications requiring materials that can stretch under load without breaking.

Figure 15-7. Material Selection Chart – Strength versus Maximum Temperature. This chart compares the strength of materials at various temperatures, aiding in the selection of materials suitable for high-temperature applications such as engines and turbines.

Figure 15-8. Material Selection Chart – Strength versus Toughness. This chart re-emphasizes the trade-off between strength and toughness for a different set of materials, providing additional insights for selecting the best materials for applications where both properties are critical.

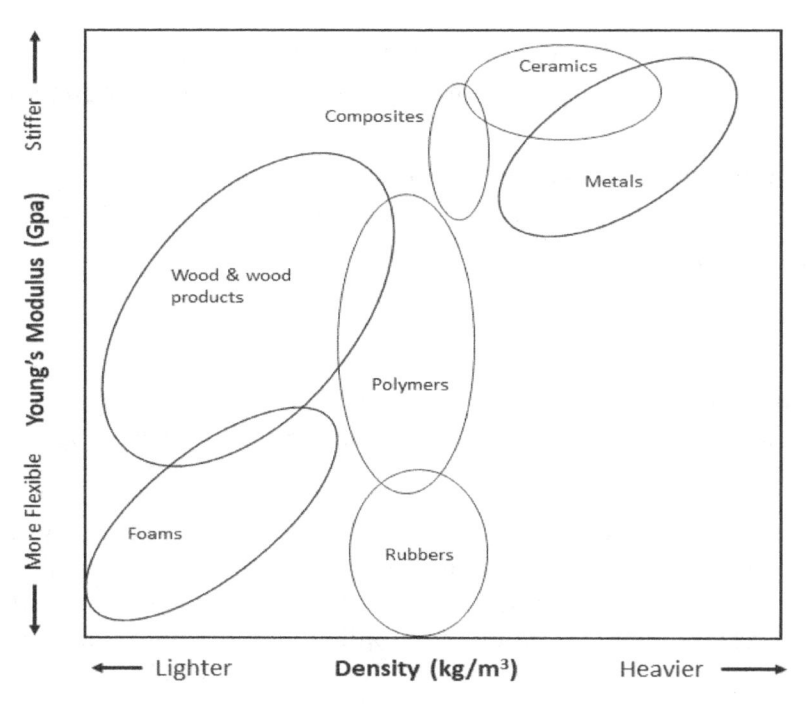

Figure 15-2. Material Selection Chart – Young's Modulus versus Density.

15.5 Materials

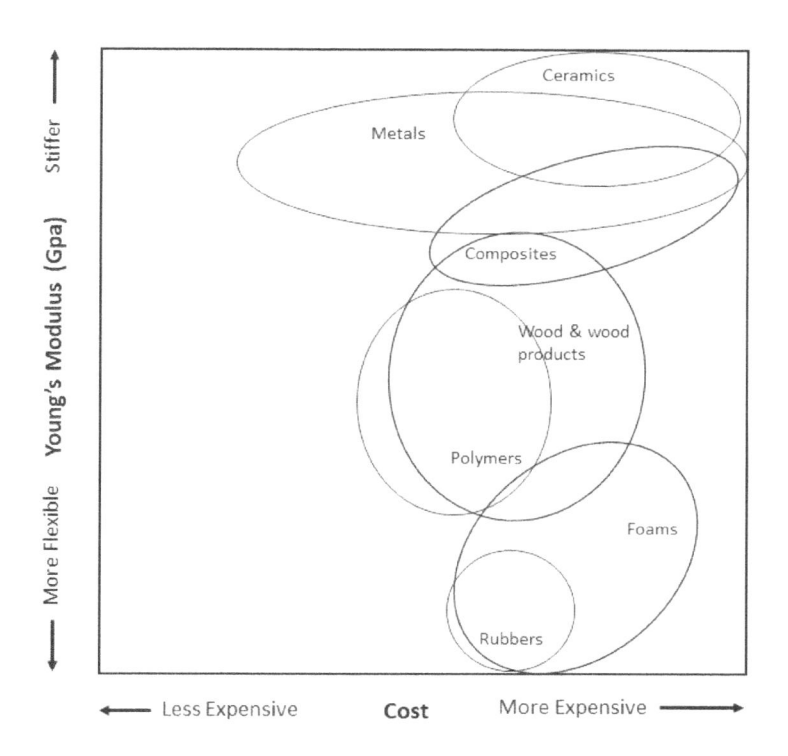

Figure 15-3. Material Selection Chart – Young's Modulus versus Cost.

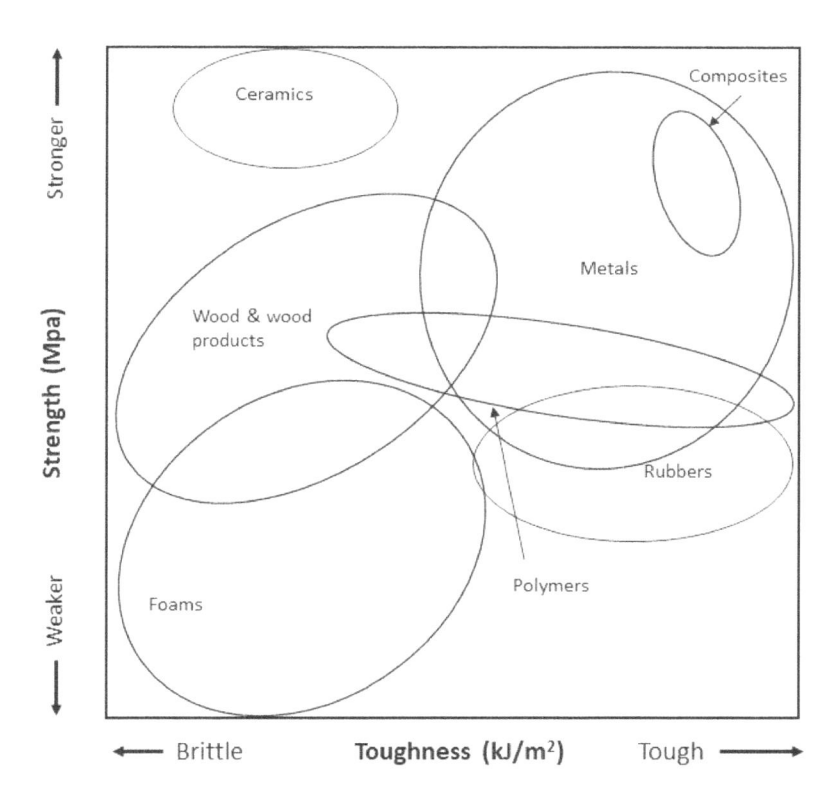

Figure 15-4. Material Selection Chart – Strength versus Toughness.

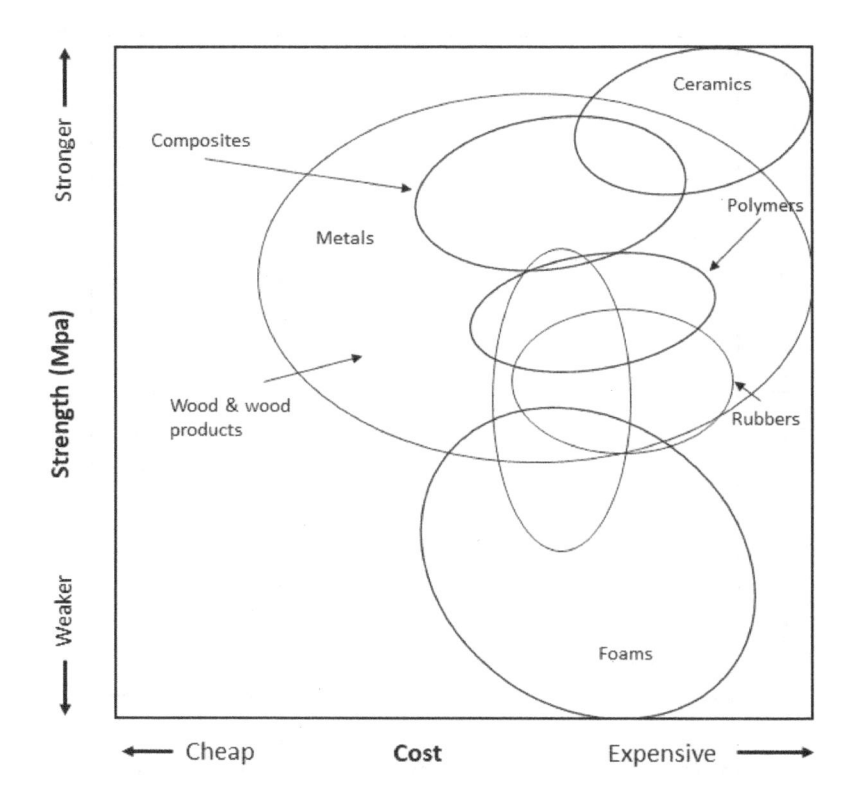

Figure 15-5. Material Selection Chart – Strength versus Cost.

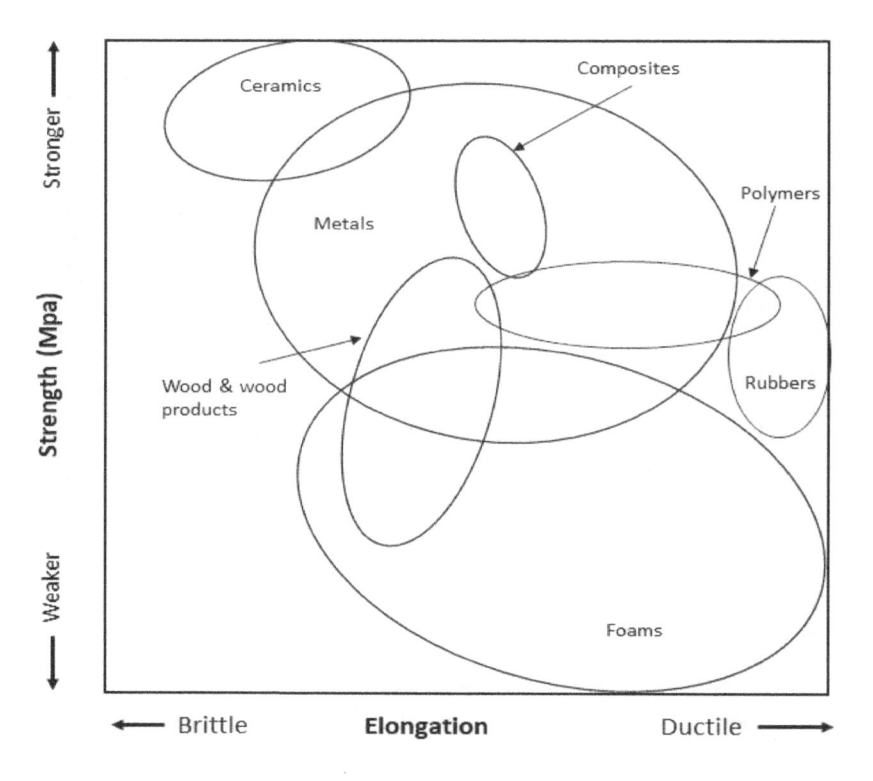

Figure 15-6. Material Selection Chart – Strength versus Elongation.

15.5 Materials

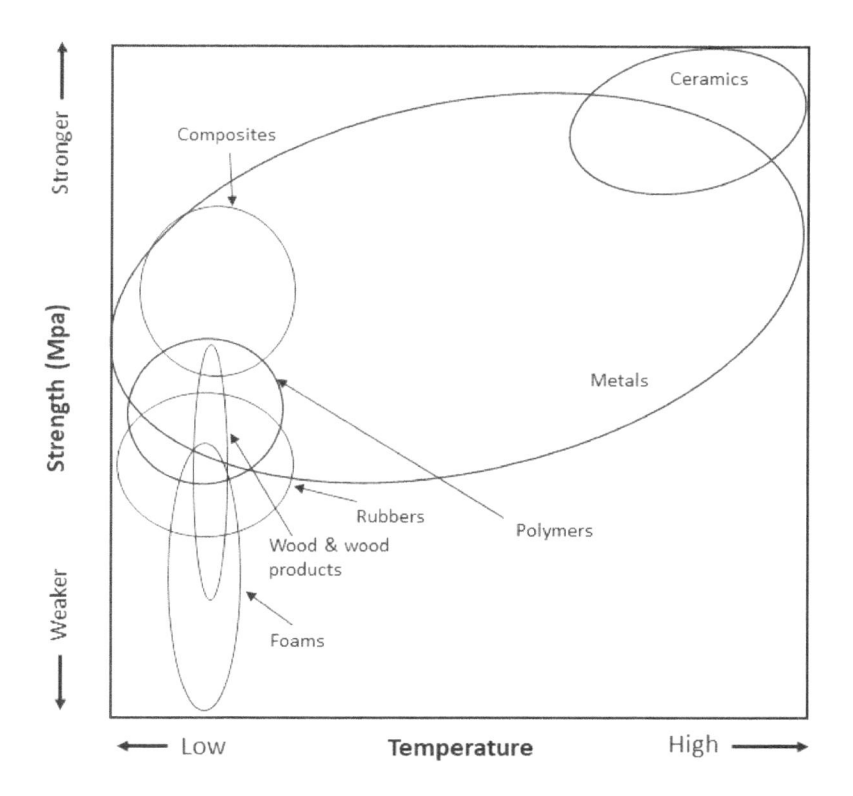

Figure 15-7. Material Selection Chart – Strength versus Maximum Temperature.

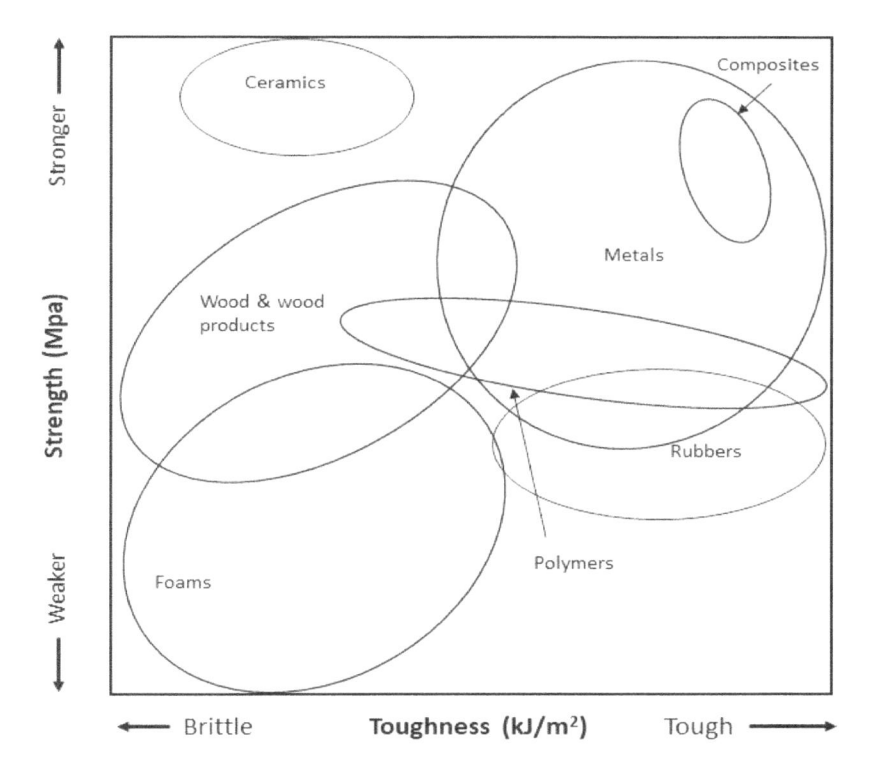

Figure 15-8. Material Selection Chart – Strength versus Toughness.

15.6 Manufacturability

Manufacturability is a measure of how effectively a product can be manufactured, considering design, application, time, cost, technology, available facilities, and skillsets. A product's manufacturability is influenced significantly by how it was designed, emphasizing the importance of designing for manufacturability (DFM). DFM focuses on reducing costs and optimizing the product's fitness for its intended purpose, ensuring it meets design specifications efficiently.

15.6.1 Key Parameters in Manufacturability

15.6.1.1 Design

- **Simplicity**: Simplify the design to minimize the number of parts and avoid complex geometries that are difficult to manufacture.
- **Standardization**: Use standard components and materials wherever possible to reduce costs and simplify procurement.
- **Tolerance and Fit**: Ensure that tolerances are realistic and achievable with the available manufacturing processes to avoid unnecessary precision that increases costs.

15.6.1.2 Application

- **Material Selection**: Choose materials that are not only suitable for the product's function but also easy to process with the available manufacturing technologies.
- **Process Suitability**: Ensure that the selected manufacturing processes are suitable for the intended application, considering factors such as strength, durability, and aesthetics.

15.6.1.3 Time

- **Lead Time**: Optimize the design to reduce lead times in manufacturing by avoiding processes that require long setup times or extended processing durations.
- **Parallel Processing**: Design for processes that can be carried out simultaneously to reduce overall production time.

15.6.1.4 Cost

- **Material Costs**: Select cost-effective materials that meet the design requirements without excessive expense.
- **Labor Costs**: Design for ease of assembly to minimize labor costs, considering automation where feasible.
- **Operational Costs**: Consider the operational costs associated with different manufacturing methods, including energy consumption and maintenance.

15.6.1.5 Technology

- **Advanced Manufacturing**: Utilize advanced manufacturing technologies like CNC machining, laser cutting, and additive manufacturing (3D printing) to improve precision and reduce waste.
- **Digital Manufacturing**: Implement digital manufacturing techniques such as CAD/CAM to streamline the design-to-manufacturing process.

15.6.1.6 Facilities and Equipment

- **Machine Shop**: Leverage the capabilities of the machine shop, including milling, turning, drilling, and grinding, to fabricate parts accurately.
- **3D Printing**: Use 3D printing for prototyping and producing complex geometries that are difficult or impossible with traditional manufacturing methods.

15.6.1.7 Skillsets

- **Team Expertise**: Assess the skillsets of the design team members and align the manufacturing methods with their expertise to maximize efficiency and minimize errors.
- **Training and Development**: Provide necessary training to team members to enhance their capabilities with advanced manufacturing techniques.

15.6.2 Methods of Manufacturing

15.6.2.1 Subtractive Manufacturing

- **Machining**: Processes like milling, turning, and drilling involve removing material from a workpiece to achieve the desired shape. These methods are precise and suitable for creating parts with tight tolerances.
- **Laser Cutting**: A high-precision method for cutting and engraving materials like metals, plastics, and composites, suitable for complex shapes and fine details.

15.6.2.2 Additive Manufacturing

- **3D Printing**: Builds parts layer by layer from materials such as plastics, metals, and composites. This method is ideal for rapid prototyping and producing intricate designs with minimal material waste.
- **Selective Laser Sintering (SLS)**: A type of 3D printing that uses a laser to sinter powdered material, suitable for producing durable and complex parts.

15.6.2.3 Forming and Shaping

- **Injection Molding**: Used for mass-producing plastic parts with high precision and repeatability. It involves injecting molten plastic into a mold cavity.

- **Casting**: Involves pouring liquid material into a mold where it solidifies. Suitable for metals and certain plastics, casting is used for parts with complex geometries.

15.6.2.4 Joining and Assembly

- **Welding**: Joins metal parts by melting them together, suitable for structural applications requiring strong joints.
- **Adhesive Bonding**: Uses adhesives to join parts, applicable for a variety of materials including metals, plastics, and composites.
- **Fastening**: Involves using screws, bolts, and rivets to assemble parts, allowing for easy disassembly and maintenance.

15.6.2.5 Surface Treatment

- **Coating**: Applies a layer of material to protect or enhance the surface properties of a part. Common coatings include paint, powder coating, and anodizing.
- **Heat Treatment**: Alters the properties of metals through controlled heating and cooling, improving hardness, strength, and wear resistance.
- **Designing for Manufacturability (DFM)**

15.6.3 Early Design Considerations

- **Manufacturing Input**: Involve manufacturing engineers early in the design process to ensure the design is manufacturable with existing processes and equipment.
- **Prototype Testing**: Create prototypes to test and refine the design, ensuring it meets manufacturability criteria before full-scale production.
  Optimization Techniques
- **Modular Design**: Design products with interchangeable modules to simplify manufacturing and assembly.
- **Design for Assembly (DFA)**: Optimize the design to make assembly processes faster, easier, and less error-prone.
- **Design for Cost (DFC)**: Focus on reducing overall costs by considering material, manufacturing, and operational expenses during the design phase.
  Quality Control
- **Tolerances and Specifications**: Clearly define tolerances and specifications to ensure consistent quality across all manufactured parts.
- **Inspection and Testing**: Implement thorough inspection and testing procedures to identify and correct defects early in the manufacturing process.
- **Case Studies and Examples**
  Example 1: Machined Metal Component
- **Material**: Aluminum 6061-T6
- **Process**: CNC Machining

- **Design Considerations**: Simplified geometry with standard hole sizes and radii to reduce machining time. Standardized tolerances to match CNC capabilities.
- **Outcome**: Reduced production time and cost, improved precision and repeatability.

Example 2: Plastic Injection Molded Part
- **Material**: ABS Plastic
- **Process**: Injection Molding
- **Design Considerations**: Uniform wall thickness to avoid warping, integrated snap-fit features for easy assembly.
- **Outcome**: High-quality, consistent parts with reduced assembly time.

Example 3: 3D Printed Prototype
- **Material**: PLA
- **Process**: Fused Deposition Modeling (FDM) 3D Printing
- **Design Considerations**: Optimized support structures for minimal post-processing, design for easy removal of supports.
- **Outcome**: Rapid prototyping with minimal material waste, enabling quick iterations and design improvements.

By understanding and applying the principles of manufacturability, design teams can create products that are not only functional and reliable but also efficient to manufacture. This holistic approach ensures that the final product meets performance requirements while being cost-effective and feasible to produce.

15.6.4 Manufacturing Process Selection

The selection of a manufacturing process is a critical step in ensuring that a design can be effectively and efficiently produced. Several factors influence this decision, including the choice of materials, the scale or size of the working model, the availability of machine tools, cost considerations, and the skill sets of the design team. This section outlines the key considerations and steps involved in selecting an appropriate manufacturing process.

15.6.4.1 Key Considerations in Manufacturing Process Selection

15.6.4.1.1 Material Choice

- **Properties and Suitability**: Different materials have varying properties such as strength, flexibility, machinability, and cost. For example, steel offers high strength and durability but is heavy and may require specialized welding and machining skills.
- **Compatibility with Processes**: Some materials are more suitable for

certain manufacturing processes. For example, metals can be cast, forged, or machined, while plastics are often molded or 3D printed.

15.6.4.1.2 Working Model Scale or Size

- **Prototype vs. Full-Scale**: Deciding whether to build a full-scale prototype or a scaled-down version depends on factors such as budget, time, and the need for functional testing under realistic conditions.
- **Handling and Transportation**: Larger models may require more complex handling and transportation arrangements, impacting the choice of manufacturing process.

15.6.4.1.3 Availability of Machine Tools

- **In-House Capabilities**: Assess the available machine tools within the team's facilities, such as CNC machines, lathes, 3D printers, and welding equipment.
- **External Resources**: Identify external resources that can be utilized if certain capabilities are not available in-house. This includes partnerships with local machine shops, fabrication facilities, or service bureaus.

15.6.4.1.4 Cost Considerations

- **Material Costs**: The cost of raw materials varies significantly. Metals like titanium are more expensive than steel or aluminum, while certain polymers may offer cost advantages for specific applications.
- **Processing Costs**: Different manufacturing processes have varying costs associated with them. For example, CNC machining may be more expensive than injection molding for high-volume production.
- **Labor Costs**: Skilled labor costs must be factored in, especially for processes that require specialized expertise such as welding or precision machining.

15.6.4.1.5 Skill Sets

- **Team Expertise**: Evaluate the skills of the design team. If the team lacks expertise in a particular process, additional training or outsourcing may be necessary.
- **Ease of Learning**: Some processes are easier to learn and implement than others. For example, 3D printing is generally more accessible to beginners compared to complex machining operations.

15.6.4.2 Comparing Manufacturing Alternatives

To select the optimal manufacturing method, it is necessary to compare

15.6 Manufacturability

alternatives based on a structured analysis. This involves outlining and modeling the entire manufacturing process for each alternative and estimating time, cost, materials, and other requirements.

15.6.4.2.1 Steps for Comparing Alternatives:

1. **Identify Alternatives**
   - List all feasible manufacturing processes for the design, considering material compatibility and design requirements.
   - Examples include CNC machining, injection molding, 3D printing, casting, and welding.

2. **Outline the Process Steps**
   - Detail each step involved in the manufacturing process for each alternative.
   - Include preparation, machining or forming, assembly, finishing, and inspection.

3. **Estimate Time and Cost**
   - Calculate the time required for each step, considering setup times, processing times, and any necessary post-processing.
   - Estimate the cost associated with each step, including material costs, machine operation costs, and labor costs.

4. **Evaluate Material and Resource Requirements**
   - Determine the quantity and type of materials needed for each process.
   - Identify any specialized tools or equipment required and their availability.

5. **Assess Feasibility and Risks**
   - Evaluate the feasibility of each alternative based on the team's capabilities and available resources.
   - Identify potential risks and challenges associated with each process, such as technical difficulties, supply chain issues, or quality control concerns.

6. **Create Scenarios and Simulations**
   - Develop scenarios for different approaches to building the design, considering best-case, worst-case, and most likely outcomes.
   - Use simulations or digital modeling tools to visualize and test the manufacturing process.

7. **Select the Optimal Method**
   - Compare the alternatives based on the analysis and select the method that offers the best balance of cost, time, quality, and

- feasibility.
  - Justify the selection with a detailed project plan and documented analysis.

15.6.5 Creating a Project Plan for Manufacturing

A comprehensive project plan for manufacturing is essential to ensure that all steps are considered and properly coordinated. The plan should include timelines, resource allocation, and detailed task lists.

15.6.5.1 Components of a Manufacturing Project Plan:

1. **Project Objectives**
   - Define the goals and objectives of the manufacturing phase, including key deliverables and success criteria.
2. **Task Breakdown**
   - Break down the manufacturing process into discrete tasks, each with a clear description and assigned responsibility.
   - Include tasks such as material procurement, machining, assembly, quality control, and finishing.
3. **Timeline and Milestones**
   - Develop a timeline that includes all tasks, with start and end dates, and dependencies.
   - Identify critical milestones, such as completion of major components, assembly stages, and final inspection.
4. **Resource Allocation**
   - Allocate resources, including personnel, equipment, and materials, to each task.
   - Ensure that all necessary resources are available when needed to avoid delays.
5. **Risk Management**
   - Identify potential risks and develop mitigation strategies.
   - Include contingency plans for dealing with unexpected issues such as equipment failure or supply chain disruptions.
6. **Monitoring and Reporting**
   - Establish a system for monitoring progress and reporting on the status of the manufacturing process.
   - Include regular check-ins, progress reports, and updates to stakeholders.

By following these steps and considerations, the design team can effectively select a manufacturing process that aligns with their project requirements and constraints.

15.6.6 Design for Manufacturing and Assembly (DFMA)

Design for Manufacturing (DFM) and Design for Assembly (DFA) are critical methodologies aimed at optimizing the design process to make product manufacturing and assembly faster, better, and cheaper. DFMA involves refining and modifying the product design to enhance manufacturability, reduce costs, and improve overall product quality. This section discusses the principles and strategies involved in DFMA, particularly in the context of capstone design projects, where creating a working model efficiently is paramount.

15.6.6.1 Principles of DFMA

15.6.6.1.1 Parts Reduction

- **Simplifying Design**: Reducing the number of parts in a product simplifies the design, assembly, and reduces costs. Fewer parts mean fewer components to procure, handle, and store. This reduction in complexity also minimizes potential points of failure and maintenance requirements.
- **Combining Parts**: If parts do not need to move relative to each other, they can be combined into a single part. This eliminates the need for additional fasteners or adhesives and reduces assembly time. For example, a flanged bolt combines a washer and a bolt, reducing installation steps.

15.6.6.1.2 Modular Design

- **Standardization**: Utilizing standardized modules or off-the-shelf components can significantly cut costs and streamline the manufacturing process. Modular design allows for easier testing, maintenance, and upgrades by enabling the independent replacement or enhancement of modules.
- **Reuse of Existing Modules**: Leveraging modules from past projects or similar designs can save time and resources. It simplifies procurement, as the modules are readily available and have known performance characteristics.

15.6.6.1.3 Standard Parts and Components

- **Off-the-Shelf Solutions**: Standard parts are typically robust, reliable, and have been tested extensively. Using these parts reduces design and development time and lowers costs compared to custom parts.
- **Multipurpose Components**: Designing components to serve multiple

functions can further reduce part count. For instance, using an aluminum enclosure for both structural support and heat dissipation combines two functions into a single part, enhancing design efficiency and functionality.

15.6.6.1.4 Optimizing Manufacturing and Fabrication

- **Finishing Operations**: Minimizing finishing operations like polishing and painting can reduce time and costs. Designs should aim to achieve functional and aesthetic requirements with minimal additional processing.
- **Tolerancing**: Proper tolerancing is crucial. Tight tolerances increase machining costs and time, while loose tolerances may compromise product quality. An optimal balance must be found through iterative testing and refinement.

15.6.6.1.5 Fasteners and Assembly Techniques

- **Reducing Fasteners**: Fasteners add complexity, time, and cost to the assembly process. Where possible, redesign parts to use snap fits, tabs, or other fastening methods that do not require additional components or tools.
- **Minimizing Variety**: If fasteners are necessary, standardize them to reduce assembly errors and streamline the process. Fewer types of fasteners mean simpler procurement and assembly procedures.

15.6.6.2 Implementing DFMA in Capstone Projects

15.6.6.2.1 Early Design and Iterative Refinement

- **Incorporating DFMA Early**: Integrate DFMA principles early in the design phase to identify potential manufacturing and assembly issues. Early consideration of manufacturability can prevent costly redesigns later.
- **Prototyping and Testing**: Use iterative prototyping to test and refine the design. Each iteration should incorporate learnings from the previous one, focusing on improving manufacturability and assembly ease.

15.6.6.2.2 Detailed Project Planning

- **Manufacturing Analysis**: Conduct a thorough analysis of manufacturing methods, considering the availability of tools, skills, and resources. Compare alternatives to determine the most efficient and cost-effective approach.
- **Project Plan Development**: Create a detailed project plan that outlines all manufacturing steps, timelines, and resource allocations. This plan should

include contingencies for potential issues and realistic estimates of time and cost.

15.6.6.2.3 Skillset and Resource Optimization

- **Leveraging Team Skills**: Match the manufacturing processes with the skills of the design team. Ensure that team members are adequately trained or consider outsourcing tasks that require specialized expertise.
- **Utilizing Available Resources**: Maximize the use of available resources, such as machine shops, 3D printers, and fabrication tools. Plan and schedule these resources efficiently to avoid delays.

15.6.6.2.4 Collaboration and Communication

- **Stakeholder Involvement**: Involve all stakeholders, including sponsors, mentors, and professors, in the design review process. Their feedback can provide valuable insights into manufacturability and assembly considerations.
- **Regular Updates and Reviews**: Hold regular project reviews to monitor progress, address issues, and make necessary adjustments. Effective communication ensures that the project stays on track and meets its objectives.

15.6.6.3 Case Study: Implementing DFMA in a Capstone Project

15.6.6.3.1 Project Overview

A capstone team is tasked with designing a portable medical device. The initial prototype consists of numerous parts, including custom machined components, off-the-shelf electronics, and various fasteners. The team applies DFMA principles to optimize the design.

15.6.6.3.2 DFMA Application

- **Parts Reduction**: The team identifies several static components that can be combined, reducing the part count by 30%. This simplification also eliminates the need for multiple fasteners.
- **Modular Design**: By standardizing the electronic module, the team ensures easy replacement and upgrade. The module can be tested independently, simplifying the overall testing process.
- **Standard Components**: Off-the-shelf enclosures and connectors are used, reducing custom machining. The team selects components that serve multiple functions, such as using a single part for both structural support and heat shielding.

- **Optimized Manufacturing**: The design is refined to minimize tight tolerances, reducing machining costs. Finishing operations are limited to essential processes, cutting down on time and expense.
- **Reduced Fasteners**: The assembly is redesigned to use snap fits and tabs where possible, significantly lowering the number of fasteners. The remaining fasteners are standardized to two types, simplifying procurement and assembly.

15.6.6.3.3 Outcome

The application of DFMA principles results in a 20% reduction in manufacturing time and a 15% cost savings. The simplified design is easier to assemble and maintain, enhancing the overall product quality and reliability.

By applying DFMA principles, capstone design teams can achieve significant improvements in manufacturability and assembly efficiency. This structured approach ensures that designs are not only innovative but also practical and cost-effective to produce, ultimately leading to successful project outcomes.

15.6.7 Mistake Proofing

Mistakes can be costly in product or process development, often leading to failures that can result in dangerous situations, harm to people, or damage to assets. Mistake-proofing, or poka-yoke (a Japanese term), is a method aimed at preventing errors from occurring or making them obvious when they do. These techniques are crucial in ensuring the reliability and safety of a product or process by minimizing human errors during the design, assembly, execution, or usage phases.

15.6.7.1 Understanding Mistake Proofing

15.6.7.1.1 Origins and Philosophy

Mistake-proofing originated in Japan as part of the Toyota Production System. The philosophy behind poka-yoke is to design processes and products in such a way that it is impossible to make mistakes or, if mistakes are made, they are immediately noticeable and can be corrected before causing harm. The approach emphasizes proactive measures over reactive ones, aiming to eliminate the causes of defects rather than inspecting and fixing them after they occur.

15.6.7.2 Key Concepts in Mistake Proofing

15.6.7.2.1 Cause-and-Effect Analysis

A cause-and-effect analysis helps identify the root cause of failures due to errors.

This analysis typically involves:

- **Brainstorming Potential Errors**: Identify all possible errors that could occur at each step of the process.
- **Root Cause Identification**: Use tools like fishbone diagrams (Ishikawa diagrams) to trace errors back to their root causes.
- **Error Prevention Strategies**: Develop strategies to eliminate the root causes or make errors immediately detectable.

15.6.7.2.2 Process Flow Chart

Creating a process flow chart is an essential step in mistake-proofing. This visual representation of the process helps identify where human errors are likely to occur. The flow chart should:

- **Detail Each Step**: Break down the process into individual steps.
- **Identify Error-Prone Steps**: Highlight steps where human intervention could lead to mistakes.
- **Analyze for Improvement**: Review each step to find ways to redesign the process to prevent errors.

15.6.7.2.3 Error Prevention Techniques

There are several techniques for preventing errors:

- **Design Modifications**: Change the design to eliminate the possibility of errors. This could involve physical changes that prevent incorrect actions.
- **Error Detection**: Implement mechanisms that detect errors immediately when they occur.
- **Automation**: Use automation to eliminate error-prone human actions.
- **User Training and Instructions**: Provide clear instructions and training to ensure correct operation.

15.6.7.2.4 Examples of Mistake Proofing

15.6.7.2.4.1 USB Connectors

One of the most familiar examples of mistake-proofing is seen in USB connectors:
- **USB-A Connectors**: These were designed with a block tab that only allows the connector to be inserted in one orientation, ensuring the correct electrical connection. However, users often attempt to insert it incorrectly and need to flip it, which can be frustrating.

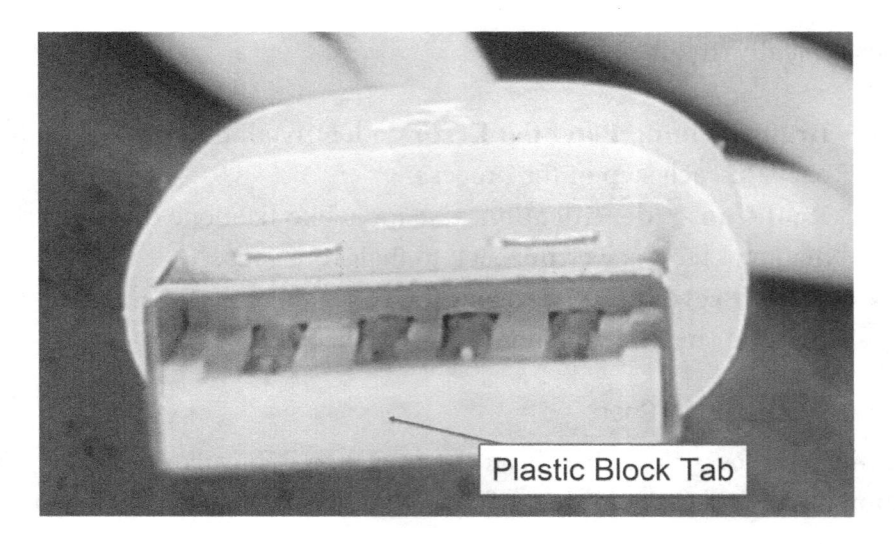

Figure 15-9. Poka-yoke example USB-A connector.

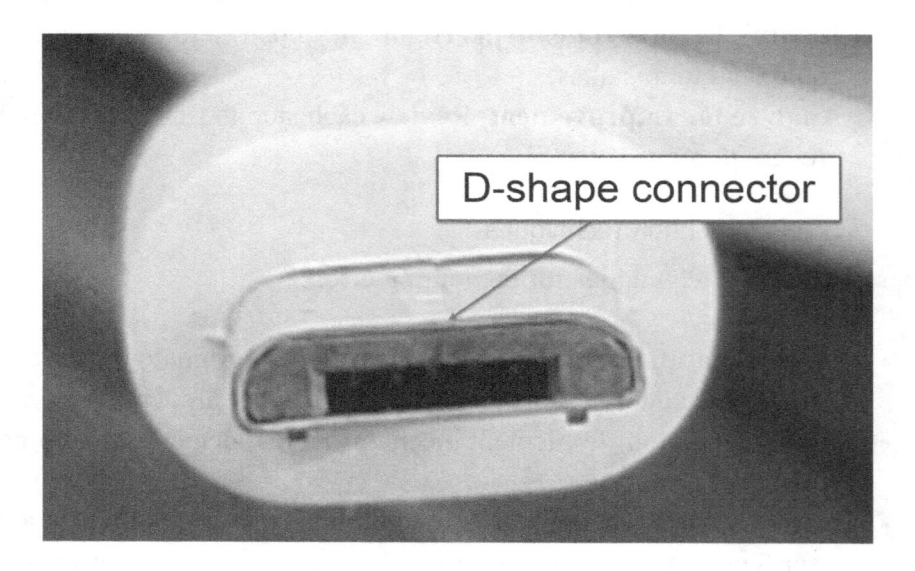

Figure 15-10. Poka-yoke example USB-micro connector.

- **USB-C Connectors**: An improved design, USB-C connectors have symmetrical contacts on both sides, allowing the plug to be inserted in either orientation without issues. This design eliminates the need for users to guess the correct way to insert the plug, showcasing a successful poka-yoke implementation.

15.6 Manufacturability

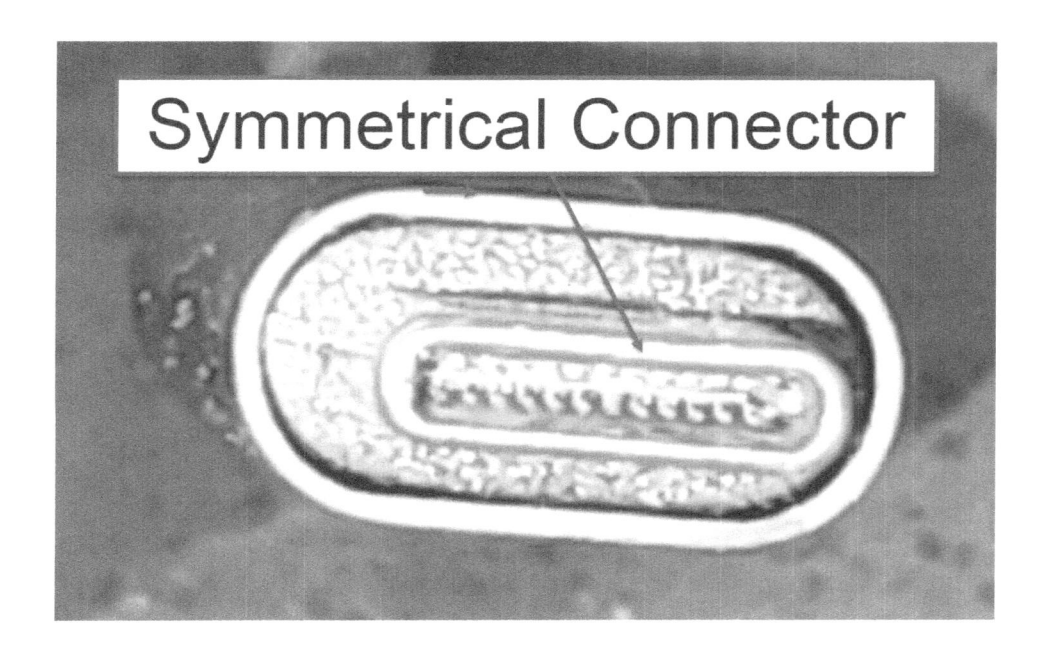

Figure 15-11. Poka-yoke example USB-C connector.

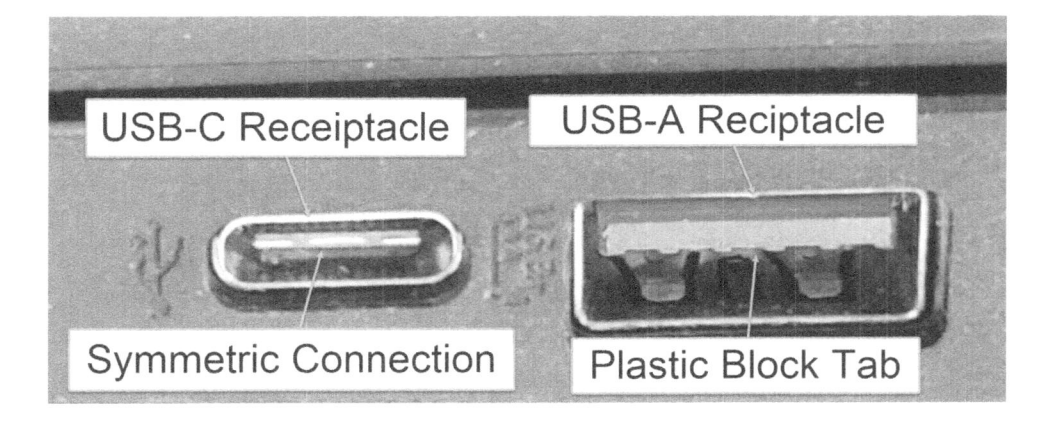

Figure 15-12. Poka-yoke example USB-C connector.

15.6.7.2.4.2 Mechanical Assemblies
- **Keyed Shafts and Gears**: These components are designed so that they can only be assembled in the correct orientation. Keys and keyways ensure that parts align properly, preventing incorrect assembly.
- **Color-Coding and Labels**: Using color-coding and clear labels for parts and connectors helps ensure that components are assembled correctly. For example, color-coded wiring in electrical assemblies reduces the risk of

incorrect connections.

15.6.7.2.5 Packaging and User Interfaces

- **Child-Resistant Packaging**: Many pharmaceutical products use child-resistant packaging that requires a specific sequence of actions (such as pressing and turning) to open. This design prevents children from accessing potentially harmful substances while allowing adults to open the package.
- **Intuitive User Interfaces**: Software and hardware interfaces designed with user-friendly features help prevent operational errors. Clear prompts, warnings, and confirmations guide users through processes, reducing the likelihood of mistakes.

15.6.7.3 Implementing Mistake Proofing in Capstone Projects

15.6.7.3.1 Early Design Phase

- **Integrate Poka-Yoke Principles Early**: Incorporate mistake-proofing techniques from the beginning of the design phase. This proactive approach helps identify and eliminate potential errors before they become embedded in the process.
- **User Scenarios and Testing**: Develop and test various user scenarios to uncover potential error points. Simulate real-world usage to identify where users might make mistakes.

15.6.7.3.2 Process Analysis

- **Create Detailed Flow Charts**: Develop comprehensive process flow charts to visualize each step of the design, manufacturing, and assembly processes. Use these charts to pinpoint where errors might occur and design safeguards against them.
- **Conduct Cause-and-Effect Analysis**: Use tools like fishbone diagrams to trace errors back to their root causes. This analysis helps in developing targeted strategies for error prevention.

15.6.7.3.3 Design and Development

- **Modular and Standardized Components**: Use modular design and standardized components to reduce complexity and the potential for errors. Modules that fit together in only one way prevent assembly mistakes.
- **Clear Documentation and Training**: Provide clear, concise documentation and training for all team members. Ensure that everyone understands the correct procedures and the reasons behind them.

15.6.7.3.4 Review and Feedback

- **Continuous Improvement**: Regularly review the design and manufacturing processes to identify new potential error points. Gather feedback from users and stakeholders to refine mistake-proofing measures.
- **Error Logs and Analysis**: Maintain logs of errors that occur during testing and assembly. Analyze these logs to identify patterns and implement corrective actions.

By integrating mistake-proofing techniques throughout the design and manufacturing process, capstone design teams can significantly reduce the likelihood of errors. This proactive approach not only enhances the quality and reliability of the final product but also improves safety and efficiency, leading to a successful project outcome.

15.6.8 3D Printing

3D printing has surged in popularity over the past decade due to critical technological advancements that have lowered costs and increased accuracy and usability. This method of additive manufacturing has become an effective way to prototype and create working models for various designs. However, successful 3D printing requires careful consideration of design parameters, materials, and best practices. This section provides an in-depth guide to understanding and utilizing 3D printing technology.

15.6.8.1 Overview of 3D Printing Technology

3D printing, or additive manufacturing, creates objects layer by layer from a digital model. This contrasts with subtractive manufacturing, where material is removed to achieve the final shape. Standard lower-priced 3D printers typically use fused filament deposition manufacturing (FFDM) technology, where a plastic filament is melted and extruded layer by layer.

15.6.8.2 History and Development

3D printing, also known as additive manufacturing, has revolutionized the manufacturing and prototyping industries by allowing the creation of three-dimensional objects from digital models. The history of 3D printing dates back to the early 1980s.

15.6.8.2.1 1980s: Inception and Early Developments

- **1981**: Hideo Kodama of the Nagoya Municipal Industrial Research Institute published the first account of a rapid prototyping system using photopolymers, a precursor to 3D printing.
- **1984**: Charles Hull invented stereolithography (SLA), a process that uses UV

lasers to cure photopolymer resin layer by layer. Hull founded 3D Systems in 1986, producing the first commercial 3D printer, the SLA-1.

15.6.8.2.2 1990s: Expansion and New Technologies

- **1991**: Stratasys introduced Fused Deposition Modeling (FDM), patented by S. Scott Crump. This technique became one of the most widely used in desktop 3D printing.
- **1992**: DTM Corporation developed Selective Laser Sintering (SLS), which uses a laser to sinter powdered material.
- **1993**: MIT patented a process called "3D printing" (3DP), which uses inkjet printer heads to deposit a liquid binder onto a powder bed.

15.6.8.2.3 2000s: Commercialization and Accessibility

- **2005**: The expiration of key patents, especially those related to FDM, led to an increase in new 3D printer manufacturers and the proliferation of consumer-grade 3D printers.
- **2009**: The RepRap project, initiated by Adrian Bowyer, aimed to create a self-replicating 3D printer. This open-source initiative significantly contributed to the popularity and affordability of 3D printing.

15.6.8.2.4 2010s: Mainstream Adoption and Technological Advancements

- **2012**: Formlabs introduced the first affordable SLA printer through a successful Kickstarter campaign.
- **2014**: HP announced its entry into the 3D printing market with Multi Jet Fusion (MJF) technology, which offers faster printing speeds and lower costs.

15.6.8.2.5 2020s: Continued Innovation and Industrial Applications

- 3D printing continues to evolve with advancements in materials, precision, and speed. Industries such as aerospace, automotive, healthcare, and construction increasingly adopt 3D printing for production and prototyping.

15.6.8.3 Ultimaker S7 and Creality K1 Max Printers

The 3D printing lab is equipped with advanced printers, including the Ultimaker S7 and the Creality K1 Max. These printers offer robust features suitable for a wide range of applications, from prototyping to small-scale manufacturing.

15.6.8.3.1 Ultimaker S7

The Ultimaker S7 is a professional-grade 3D printer known for its reliability, precision, and ease of use. It is part of Ultimaker's S-line series, designed for high-performance printing with a variety of materials.

- **Build Volume**: 330 x 240 x 300 mm, providing ample space for large or multiple parts.
- **Dual Extrusion**: Allows for printing with two different materials or colors simultaneously, enhancing versatility.
- **Heated Bed**: Ensures better adhesion and reduces warping, particularly useful for printing with materials like ABS and Nylon.
- **Filament Compatibility**: Supports a wide range of materials, including PLA, ABS, Nylon, CPE, and PVA.
- **Precision and Speed**: Layer resolution down to 20 microns and a print speed of up to 300 mm/s.
- **Touchscreen Interface**: User-friendly touchscreen for easy navigation and control.
- **Wi-Fi Connectivity**: Enables remote monitoring and printing, integrating seamlessly with Ultimaker's Cura software for slicing and printer management.

15.6.8.3.2 Creality K1 Max

The Creality K1 Max is a powerful and large-format 3D printer designed for users needing extensive build volumes and robust performance. It is known for its affordability and reliability, making it a popular choice among hobbyists and professionals alike.

- **Build Volume**: 300 x 300 x 400 mm, suitable for creating large prototypes and parts.
- **Direct Drive Extruder**: Provides better control over filament extrusion, enhancing the ability to print flexible and exotic materials.
- **Heated Bed**: Ensures good adhesion and reduces warping, essential for successful prints with various filaments.
- **Filament Compatibility**: Supports materials like PLA, PETG, TPU, ABS, and wood-filled filaments.
- **Precision and Speed**: Layer resolution down to 100 microns and a print speed of up to 180 mm/s.
- **Touchscreen Interface**: Intuitive touchscreen for straightforward operation and adjustments.
- **Automatic Bed Leveling**: Simplifies the setup process and ensures a consistent first layer for high-quality prints.

Both the Ultimaker S7 and Creality K1 Max provide the necessary features and capabilities to handle complex 3D printing tasks, making them invaluable tools in the 3D printing lab for educational, prototyping, and small-scale manufacturing purposes.

15.6.8.4 Printer Components

- **Print Head**: Moves in the x-y plane to deposit material.
- **Build Plate**: Moves in the z-direction to create layers.
- **Filament**: Typically, 1.75 mm or 2.85 mm in diameter, with a standard nozzle size of 0.4 mm.

15.6.8.5 Controller and Firmware

- **Microcontroller**: Often an Arduino controller, manages print head and bed movements.
- **Firmware**: Standard open-source firmware like Marlin controls the printer's operations.

15.6.8.6 3D Printing Process

15.6.8.6.1 Process Flow

1. **Design Creation**: Use CAD software to create a 3D model.
2. **File Export**: Export the model in stereolithography (STL) format.
3. **Slicing**: Use slicing software to convert the STL file into gcode, the printer's instruction set.
4. **Printing**: Load the g-code onto the 3D printer via SD card or USB and execute the print.

15.6.8.6.2 Design Considerations

- **Geometry**: Ensure walls are thick enough to be printable and avoid features that result in unprintable internal geometries.
- **Material Usage**: Optimize the design to minimize material waste and printing time.

15.6.8.6.3 Materials for 3D Printing

15.6.8.6.3.1 Common Filament Types

- **PLA (Polylactic Acid)**: Biodegradable and easy to print.
- **ABS (Acrylonitrile Butadiene Styrene)**: Stronger and more heat-resistant but requires a heated bed.
- **PETG (Polyethylene Terephthalate Glycol)**: Durable and flexible, good for functional parts.
- **Nylon**: Strong, flexible, and abrasion-resistant.

Material Selection Considerations
- **Mechanical Properties**: Choose materials based on the required strength, flexibility, and durability.
- **Thermal Properties**: Consider the material's melting point and thermal stability.
- **Chemical Resistance**: Select materials that are resistant to chemicals if required by the application.

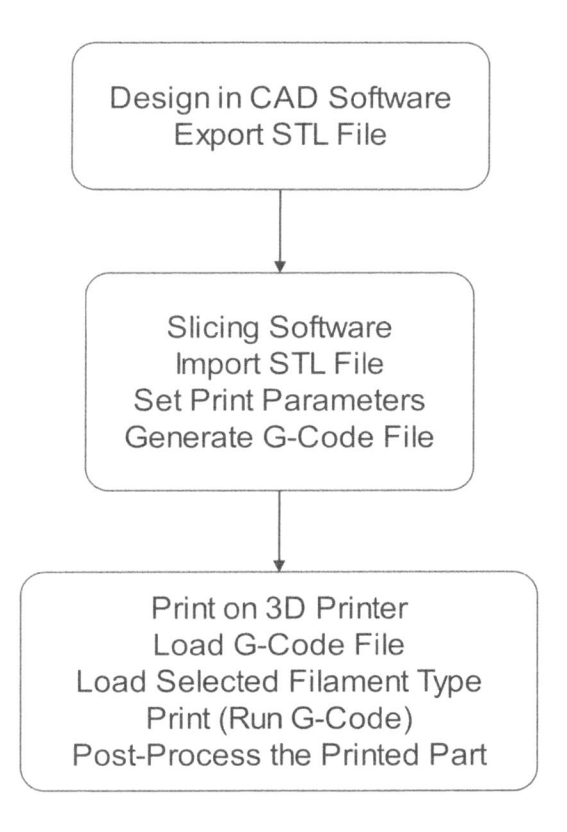

```
┌─────────────────────────┐
│  Design in CAD Software │
│     Export STL File     │
└─────────────────────────┘
              │
              ▼
┌─────────────────────────┐
│     Slicing Software    │
│     Import STL File     │
│   Set Print Parameters  │
│   Generate G-Code File  │
└─────────────────────────┘
              │
              ▼
┌───────────────────────────────┐
│      Print on 3D Printer      │
│       Load G-Code File        │
│  Load Selected Filament Type  │
│      Print (Run G-Code)       │
│  Post-Process the Printed Part│
└───────────────────────────────┘
```

Figure 15-13. 3D Printing Process.

15.6.8.6.4 Slicing Software and File Preparation

15.6.8.6.4.1 STL File Preparation

- **Conversion**: The CAD model is converted to STL format, representing geometry as a mesh of triangles.
- **Resolution**: Use the highest resolution allowed by the CAD software to ensure accuracy.
- **Mesh Repair**: Address issues like small or long triangles, non-watertight surfaces, and incorrectly oriented surface normals using tools like Meshlab or Simplify3D.

15.6.8.6.4.2 Slicing Software

- **Simplify3D**: A versatile slicer supporting many printer types, allows detailed customization of print settings.
- **Vendor-Specific Slicers**: Often provided free by printer manufacturers, optimized for their specific machines.

15.6.8.6.4.3 Slicing Parameters

- **Printer Settings**: Input the printer's characteristics, such as bed dimensions, z-travel, and nozzle size.
- **Print Settings**: Adjust parameters like layer height, print speed, and temperatures for the nozzle and bed.
- **Output**: Generate gcode to control the printer.

15.6.8.7 Best Practices for 3D Printing

15.6.8.7.1 Pre-Print Setup

- **Bed Leveling**: Ensure the build plate is level for optimal adhesion.
- **Filament Loading**: Properly load the filament and check for smooth extrusion.

Printing Tips

- **Adhesion**: Use adhesives or textured surfaces to enhance bed adhesion.
- **Supports**: Include supports for overhanging features and remove them post-print.
- **Monitoring**: Monitor the print process to catch issues early, such as filament jams or layer shifting.

15.6.8.7.2 Post-Print Processing

- **Removal**: Carefully remove the print from the bed to avoid damage.
- **Cleaning**: Trim any excess material or supports.
- **Finishing**: Sanding, painting, or other finishing techniques can be applied to improve the appearance and functionality.

15.6.8.8 Troubleshooting Common Issues

15.6.8.8.1 Layer Shifting

- **Causes**: Mechanical issues, loose belts, or obstructions.
- **Solutions**: Tighten belts, ensure smooth movement of the print head, and check for obstructions.

15.6.8.8.2 Poor Adhesion

- **Causes**: Unleveled bed, incorrect temperatures, or improper bed surface.
- **Solutions**: Relevel the bed, adjust temperatures, and use adhesives or textured surfaces.

15.6.8.8.3 Stringing

- **Causes**: High print temperatures or insufficient retraction.
- **Solutions**: Lower print temperatures and increase retraction settings.

15.6.8.8.4 Warping

- **Causes**: Uneven cooling or insufficient bed adhesion.
- **Solutions**: Use a heated bed, enclosures, and adhesives to improve adhesion.

15.6.8.9 Advanced Techniques

15.6.8.9.1 Dual Extrusion

- **Benefits**: Allows for multi-material prints or dual-color designs.
- **Considerations**: Requires a printer with dual extruders and careful calibration to avoid misalignment.

15.6.8.9.2 Large-Scale Printing

- **Techniques**: Use printers with larger build volumes or segment the model into smaller parts for assembly.
- **Challenges**: Increased print times and potential for warping or layer adhesion issues.

15.6.8.9.3 High-Resolution Printing

- **Settings**: Use smaller nozzle sizes and layer heights for finer details.
- **Trade-offs**: Longer print times and increased risk of printing errors.

By understanding and applying these principles and best practices, design teams can effectively utilize 3D printing to create accurate, functional prototypes and working models. This technology not only enhances the design process but also enables rapid iteration and testing, leading to more innovative and successful capstone projects.

15.6.9 Quality

The quality of a design-build project, whether a product or a process, depends significantly on the skill sets of the engineering students involved. Engineering programs typically emphasize theoretical knowledge and analytical skills rather than hands-on building or process engineering required for production-ready creations. However, students who have gained practical experience through co-op programs or internships can be invaluable to their teams. They bring practical insights and skills that enhance the quality of the design and its implementation. This section explores how quality can be achieved in engineering capstone projects and the importance of practical experience in the design and build phases.

15.6.9.1 Importance of Skill Sets

15.6.9.1.1 Diverse Skill Sets

- **Analytical Skills**: Engineering programs focus heavily on analytical skills, which are crucial for the initial design and problem-solving stages.
- **Practical Skills**: Practical skills in building and process engineering are often acquired outside the academic curriculum through internships, co-ops, or personal projects. These skills are essential for translating theoretical designs into functional prototypes or processes.

15.6.9.1.2 Leveraging Internships and Co-op Experiences

- **Application of Theory**: Students with co-op or internship experiences can apply theoretical knowledge in practical settings, bridging the gap between design and implementation.
- **Team Assets**: These students can share their practical insights with their teammates, enhancing the overall skill set of the team and improving the quality of the project.

15.6.9.1.3 Assessing and Matching Skills

15.6.9.1.3.1 Self-Assessment and Team Evaluation

- **Identifying Strengths and Weaknesses**: Teams should assess their combined skill sets early in the project to identify strengths and potential gaps.
- **Role Assignment**: Assign roles based on individual strengths. For instance, students with hands-on experience should lead the build or implementation phase, while those with strong analytical skills might focus on design optimization and problem-solving.

15.6.9.1.3.2 Skill Set Gaps and Mitigation

- **Training and Workshops**: If the team lacks certain skills, they should seek additional training or workshops. Many universities offer maker spaces or workshops where students can learn practical skills.
- **Mentorship**: Engage with mentors or advisors who can provide guidance and fill

knowledge gaps. This external expertise can be critical in achieving a high-quality build.

15.6.9.2 Quality in Design Implementation

15.6.9.2.1 Design for Manufacturability and Assembly (DFMA)
- **Simplified Designs**: Focus on creating designs that are easy to manufacture and assemble. This reduces the complexity and potential errors during the build phase.
- **Iterative Testing and Refinement**: Implement an iterative process where designs are continuously tested and refined based on build experiences.

15.6.9.2.2 Practical Implementation Considerations
- **Realistic Designs**: Ensure that designs are realistic and within the capabilities of the team's skills and available resources.
- **Prototype Testing**: Build and test prototypes to identify and rectify design flaws before final implementation.

15.6.9.3 Objective Quality Assessment

15.6.9.3.1 External Evaluation
- **Sponsor and Professor Review**: Involve project sponsors, capstone professors, and mentors in the evaluation process to ensure an objective assessment of quality.
- **Peer Reviews**: Conduct peer reviews within the team and with other teams to gain diverse perspectives on the quality of the design and build.

15.6.9.4 Criteria for Quality Assessment
- **Functionality**: The design should meet all functional requirements and specifications outlined at the project's inception.
- **Reliability and Durability**: The prototype or process should be reliable and durable, capable of withstanding expected operating conditions.
- **Aesthetics and Usability**: Consider the end-user experience, ensuring the design is aesthetically pleasing and user-friendly.
- **Safety and Compliance**: Ensure that the design complies with relevant safety standards and regulations.

15.6.9.5 Learning from the Build Process

15.6.9.5.1 Hands-On Experience

- **Building the Design**: Encourage students to build their design solutions rather than outsourcing. This hands-on experience is invaluable for understanding the practical implications of design choices.
- **Learning from Mistakes**: Building the prototype allows students to learn from mistakes and make necessary design adjustments, an essential part of the engineering design process.

15.6.9.5.2 Feedback and Iteration

- **Continuous Improvement**: Use feedback from testing and external evaluations to iteratively improve the design.
- **Documenting Changes**: Keep detailed records of all changes made during the build process. This documentation can help identify patterns and recurring issues that need addressing.

15.6.9.6 Case Study: Capstone Project Quality Improvement

15.6.9.6.1 Project Overview

A capstone team is tasked with designing and building an automated greenhouse system. The team consists of members with varying levels of practical experience.

15.6.9.6.2 Initial Skill Assessment

- **Identifying Gaps**: The team identifies a gap in practical electrical and mechanical assembly skills.
- **Role Assignment**: Members with relevant co-op experience take the lead in these areas, while others focus on software and design optimization.

15.6.9.6.3 DFMA Principles

- **Simplified Design**: The team designs modular components that are easy to assemble and replace.
- **Iterative Prototyping**: They build and test each module separately, identifying and fixing issues before integrating the system.

15.6.9.6.4 Objective Quality Assessment

- **External Reviews**: The team schedules regular reviews with their professor and project sponsor to assess progress and quality.
- **Peer Feedback**: They also engage in peer reviews with other capstone teams to gain additional insights.

15.6.9.6.5 Learning and Iteration

- **Hands-On Building**: All team members participate in the build process,

gaining practical experience and understanding the implications of their design choices.
- **Documenting Iterations**: The team maintains detailed documentation of all changes and improvements made, providing a valuable reference for future projects.

Quality in capstone design projects hinges on the practical skills of the engineering students and their ability to integrate theoretical knowledge with hands-on experience. By leveraging diverse skill sets, engaging in continuous learning and iteration, and involving external evaluators, student teams can achieve high-quality outcomes that meet functional, aesthetic, and safety standards. Encouraging hands-on involvement in the build process not only enhances the quality of the final product but also provides invaluable learning experiences that prepare students for their future careers in engineering.

15.6.9.7 Bill of Materials

The engineering bill of materials (BOM) is a comprehensive list created by the engineering design team that details all the assemblies, sub-assemblies, parts, components, and materials necessary to build a design. It serves as a blueprint for manufacturing, ensuring that every component required for the construction of the product is accounted for. A well-structured BOM is crucial for efficient production, cost management, and minimizing errors.

15.6.9.7.1 Importance of the Bill of Materials

15.6.9.7.1.1 Detailed Documentation

- **Assemblies and Sub-Assemblies**: The BOM breaks down the product into its major assemblies and further into sub-assemblies, ensuring a clear hierarchical structure.
- **Parts and Components**: It lists all individual parts and components, providing specifications such as dimensions, materials, and quantities.
- **Materials**: Specifies the raw materials needed, including their types and properties.

15.6.9.7.1.2 Manufacturing Preparedness

- **Procurement**: A complete BOM allows manufacturers to prepare for the necessary procurement steps, ensuring that all parts and materials are available when needed.
- **Cost Management**: Helps in finding the best pricing, quality, and quantity for parts, reducing overall manufacturing costs.

- **Supply Chain Efficiency**: Facilitates competitive procurement and shortens supply-chain lead times by providing clear and detailed requirements.

15.6.9.7.2 Quality and Completeness

- **Accuracy**: The accuracy of the BOM minimizes errors and rework, which can be costly and time-consuming.
- **Design Maturity**: Reflects the maturity and thoroughness of the design process. A well-prepared BOM indicates that the design team has considered all aspects of the product's construction.

15.6.9.7.3 Creating the Bill of Materials

15.6.9.7.3.1 Automated vs. Manual Creation

- **CAD Software**: Tools like SolidWorks can generate a BOM if the CAD models are detailed enough.
- **Manual Creation**: In some cases, the design team may manually create the BOM, ensuring that every item is accurately listed and described.

15.6.9.7.3.2 Components of a BOM

1. **Part Number**: A unique identifier for each part or component.
2. **Part Name/Description**: A brief description of the part, including its function.
3. **Quantity**: The number of each part required.
4. **Material**: Specifies the material from which the part is made.
5. **Dimensions/Specifications**: Key dimensions and specifications that define the part.
6. **Source/Vendor**: Information about where the part can be procured.
7. **Cost**: The cost per unit of the part.
8. **Assembly/Location**: Indicates where the part fits into the overall assembly.

15.6.9.7.4 Example BOM

To illustrate, let's consider a simple example of a BOM for a basic drone assembly.

15.6 Manufacturability

Table 15-1. Example Bill of Materials Table.

Part No.	Part Name/ Description	Quantity	Material	Dimensions/ Specifications	Source/ Vendor	Cost (per unit)	Assembly/ Location
001	Frame	1	Carbon Fiber	450mm x 450mm	DroneParts Inc.	$50	Main Structure
002	Motor	4	Metal/Plastic	920KV	Motors R Us	$20	Each Arm (4 total)
003	Propeller	4	Plastic	10x4.5 inch	Propeller World	$5	Each Motor (4 total)
004	ESC (Electronic Speed Controller)	4	Metal/Plastic	30A	Electronics Hub	$15	Each Motor (4 total)
005	Flight Controller	1	PCB/Plastic	Naze32	FlyControl Co.	$40	Center Frame
006	Battery	1	LiPo	2200mAh, 3S	PowerUp Batteries	$25	Battery Compartment
007	Battery Straps	2	Velcro	20mm x 300mm	Straps Galore	$2	Battery Compartment
008	Receiver	1	PCB/Plastic	2.4GHz	RC Supplies	$15	Near Flight Controller
009	Transmitter	1	PCB/Plastic	2.4GHz	RC Supplies	$30	Remote Control
010	Power Distribution Board	1	PCB	50A	Power Boards Inc.	$10	Center Frame
011	Screws and Fasteners	50	Metal	M3 x 10mm	Fastener World	$0.10	Throughout
012	GPS Module	1	PCB/Plastic	NEO-M8N	GPS Tech	$20	Frame
013	Landing Gear	4	Plastic	10cm height	DroneParts Inc.	$8	Each Arm (4 total)

15.6.9.7.4.1 Example Explanation

- **Part Number**: Each part has a unique number (e.g., 001 for the frame, 002 for the motor).
- **Part Name/Description**: Clear and concise description (e.g., Motor, Propeller).
- **Quantity**: Specifies how many of each part are needed (e.g., 4 motors, 1 frame).
- **Material**: The material used for the part (e.g., Carbon Fiber, Plastic).
- **Dimensions/Specifications**: Key details such as size or rating (e.g.,

920KV for motors, 10x4.5 inch for propellers).
- **Source/Vendor**: Where the part can be purchased (e.g., DroneParts Inc., Motors R Us).
- **Cost**: Cost per unit to aid in budgeting (e.g., $50 for the frame).
- **Assembly/Location**: Where the part is used in the assembly (e.g., Main Structure for the frame, Each Arm for motors).

15.6.9.7.5 Benefits of a Detailed BOM

15.6.9.7.5.1 Efficiency in Manufacturing

- **Streamlined Procurement**: Ensures all parts are sourced timely, avoiding delays.
- **Cost Management**: Helps in budgeting and finding cost-effective suppliers.

15.6.9.7.5.2 Quality Control

- **Minimizing Errors**: Reduces the risk of assembly errors by providing detailed information.
- **Consistent Quality**: Ensures that the correct parts and materials are used, maintaining product quality.

15.6.9.7.5.3 Project Management

- **Project Planning**: Facilitates detailed project planning and scheduling by outlining all required components.
- **Resource Allocation**: Helps in the efficient allocation of resources, ensuring everything needed is available.

15.6.9.7.5.4 Educational Value

- **Learning Tool**: Creating a BOM helps students understand the importance of detail and accuracy in engineering projects.
- **Skill Development**: Enhances skills in documentation, organization, and project management.

The Bill of Materials (BOM) is an essential document in the design and manufacturing process, ensuring that all parts, components, and materials required for a product are detailed and organized. A comprehensive and accurate BOM facilitates efficient manufacturing, cost management, and quality control. For engineering students, the process of creating a BOM provides valuable insights into the complexities of production and the importance of meticulous planning and documentation. By mastering the creation of a detailed BOM, students can significantly enhance the quality and success of their design projects.

15.7 Procurement

The procurement process is a critical aspect of engineering capstone projects, encompassing the acquisition of all items listed in the Bill of Materials (BOM) necessary for design, prototyping, building, testing, and redesign. Effective procurement ensures that student design teams have the materials and components they need to successfully complete their projects on time and within budget.

15.7.1 University Procurement Policies

15.7.1.1 Variability of Policies

Each university or school has its own set of policies and procedures for purchasing items. These policies can vary widely, especially between private institutions and state universities. State universities often have more bureaucratic purchasing policies due to regulatory requirements.

15.7.1.2 Purchasing Cards (P-cards)

Many schools issue purchasing credit cards, or P-cards, to expedite the procurement process. P-cards are particularly useful for quickly purchasing small parts and items that are typically needed for student design projects.
- **Advantages of P-cards**: They streamline the purchasing process, reduce paperwork, and allow for faster acquisition of materials.
- **Limitations**: There may be spending limits and restrictions on the types of items that can be purchased using a P-card.

15.7.2 Budget Management

15.7.2.1 Project Budget

The budget for a capstone project can range from a few hundred dollars to several thousand, depending on the project's scope. It is essential for teams to manage their budgets carefully to ensure that they have the necessary funds to complete their projects.

15.7.3 Internal Process for Procurement

An organized internal process for managing procurement needs is crucial for the capstone course, especially for large classes with numerous teams. This process should include:
- **Purchase Requests**: Teams should submit detailed purchase requests that include item descriptions, quantities, costs, and vendor information.
- **Tracking and Documentation**: Keep track of all purchase requests and document the status of each request to ensure timely approval and processing.

- **Communication**: Regularly communicate with the individuals responsible for reviewing and approving purchase requests to avoid delays.

15.7.4 Steps in the Procurement Process

15.7.4.1 Step 1: Identifying Needs

- **Bill of Materials**: Use the BOM to identify all the items that need to be procured.
- **Prioritization**: Determine the priority of each item based on the project timeline and criticality to the project's success.

15.7.4.2 Step 2: Vendor Selection

- **Approved Vendors**: Use university-approved vendors when possible to streamline the procurement process.
- **Comparative Shopping**: Compare prices and terms from multiple vendors to ensure the best value for the required items.
- **Supplier Reliability**: Consider the reliability and reputation of vendors to avoid issues with delivery and quality.

15.7.4.3 Step 3: Purchase Request Submission

- **Detailed Requests**: Submit detailed purchase requests that include item specifications, quantities, costs, vendor information, and justification for the purchase.
- **Approval Process**: Follow the university's approval process, which may include reviews by professors, department heads, or procurement officers.

15.7.4.4 Step 4: Order Placement and Tracking

- **Placing Orders**: Once approved, place orders with the selected vendors. Use the P-card if applicable.
- **Order Tracking**: Track orders to ensure timely delivery. Follow up with vendors as needed to address any issues or delays.

15.7.4.5 Step 5: Receiving and Verification

- **Receiving Goods**: Upon receipt of the ordered items, verify that all items match the purchase order in terms of specifications and quantities.
- **Inspection**: Inspect items for any damage or defects. Report any issues immediately to the vendor for resolution.

15.7.4.6 Step 6: Documentation and Reporting

- **Record Keeping**: Maintain detailed records of all purchases, including receipts, invoices, and delivery notes.
- **Budget Tracking**: Regularly update budget records to reflect expenditures and ensure that the project remains within budget.
- **Reporting**: Provide periodic reports to the project sponsor, professor, and other stakeholders on the status of procurement and budget.

15.7.5 Best Practices for Effective Procurement

15.7.5.1 Organized Documentation

- **Centralized System**: Use a centralized system for tracking purchase requests, orders, and expenditures. This can be an Excel spreadsheet, a dedicated procurement software, or an online project management tool.
- **Detailed Records**: Keep detailed records of all transactions to facilitate budget tracking and auditing.

15.7.5.2 Proactive Communication

- **Regular Updates**: Provide regular updates to team members and stakeholders on the status of procurement activities.
- **Timely Reminders**: Send timely reminders to individuals responsible for approvals and processing to avoid delays.

15.7.5.3 Vendor Relationships

- **Building Relationships**: Establish good relationships with key vendors to negotiate better terms and ensure priority service.
- **Feedback Loop**: Provide feedback to vendors on their performance to help improve future transactions.

15.7.5.4 Contingency Planning

- **Backup Suppliers**: Identify backup suppliers for critical items to mitigate risks associated with vendor issues or supply chain disruptions.
- **Buffer Stock**: Maintain a small buffer stock of essential items to avoid project delays due to procurement issues.

15.7.5.5 Training and Orientation

- **Procurement Training**: Provide training to team members on the university's procurement policies and procedures.
- **Orientation Sessions**: Conduct orientation sessions at the beginning of the

project to ensure everyone understands the procurement process and their roles.

15.7.6 Example of a Procurement Workflow

1. **Identify Needs**: Team identifies all items required from the BOM.
2. **Submit Purchase Request**: Team fills out a purchase request form with detailed information.
3. **Review and Approval**: Request is reviewed and approved by the professor or procurement officer.
4. **Order Placement**: Approved request is used to place an order with the vendor using a P-card or purchase order.
5. **Track and Receive**: Team tracks the order and verifies the received items.
6. **Update Records**: All transactions are documented, and budget records are updated.

By following these detailed procurement steps and best practices, student design teams can effectively manage their procurement needs, ensuring timely acquisition of materials and components necessary for the successful completion of their capstone projects. This process not only helps in maintaining project timelines and budgets but also provides valuable real-world experience in managing the procurement aspects of engineering projects.

15.8 Assignments

Assignment 15-1: Capstone Team Assignment: Project Plan for Building a Working Model

- **Objective:**

Teams will create a detailed project plan to build a working model of their design during the first month of the spring semester. The model should be ready for comprehensive test engineering activities by the end of the month. The assignment focuses on rapid planning and execution to meet this critical milestone.

- **Assignment Overview:**

Each team will:
  - **Develop a project plan** outlining tasks, timelines, resources, and risks.
  - **Create a build plan** detailing the specific steps and materials needed for construction.
  - **Complete the build within 4 weeks** and prepare for detailed testing.

- **Step-by-Step Guidance:**
  1. **Review Design and Goals**

     Objective: Start by revisiting your fall semester design to understand the requirements for the working model.

     Tasks:
     - **Refine Design**: Review your proof of concept (POC) prototype. Identify any weaknesses, improvements, or changes needed for the working model.
     - **Define Key Functions**: Clearly articulate what the working model must achieve. Include functional usability and any specific performance metrics.

     Deliverable: An updated design report highlighting critical functionalities and any revisions based on feedback from the prototype stage.

  2. **Develop a Project Plan**

     Objective: Break down the construction process into manageable tasks with clear milestones.

     Tasks:
     - **List Major Phases**: Include tasks such as material procurement, fabrication, assembly, testing, and documentation.
     - **Assign Responsibilities**: Assign team members to each task. Ensure that roles for engineering, procurement, assembly, and testing are clearly defined.
     - **Establish Milestones**: Set clear deadlines for each phase. Example

milestones: material procurement complete by Day 5, assembly complete by Day 20.
- **Create a Gantt Chart**: Develop a timeline to track progress.

Deliverable: A project management document that includes a Gantt chart with tasks, deadlines, and team member responsibilities.

3. Resource Planning

Objective: Ensure that all materials, tools, and facilities are available when needed.

Tasks:
- **Create a Bill of Materials (BOM)**: List every component required, including quantities, specifications, and suppliers. Prioritize items based on availability and lead times.
- **Identify Facility Needs**: Secure access to machine shops, 3D printers, or other facilities. Schedule use of shared resources well in advance.
- **Budgeting**: Ensure all materials fit within your budget, keeping in mind any contingency funds for unforeseen costs.

Deliverable: A finalized BOM and a detailed budget estimate for materials, equipment, and resources.

4. Build Plan

Objective: Outline the detailed steps necessary to construct the working model.

Tasks:
- **Define Assembly Steps**: Break down the build into clear, step-by-step tasks. Detail what needs to be assembled, fabricated, or integrated.
- **Prototyping**: If necessary, conduct small-scale tests or create sub-assemblies before the full build to validate critical parts.
- **Integration and Testing**: After assembly, integrate all subsystems and test for functionality.

Deliverable: A detailed build plan, including assembly instructions, testing plans, and integration procedures.

5. Risk Management

Objective: Anticipate potential challenges and develop strategies to mitigate risks.

Tasks:
- **Identify Risks**: Highlight risks such as material delays, equipment failures, or design issues.
- **Develop Mitigation Plans**: Create backup plans for each identified risk. For example, establish alternate suppliers for critical

15.8 Assignments

components.

Deliverable: A risk management plan with identified risks and corresponding mitigation strategies.

6. Testing and Validation

Objective: Ensure that the model is functional and ready for test engineering activities.

Tasks:

- **Initial Testing**: Conduct initial tests to ensure that key functionalities are working. Test the subsystems individually before full integration.
- **Final Testing Preparation**: Prepare for comprehensive testing, ensuring that all documentation and test plans are ready.

Deliverable: A brief test report showing initial test results and confirming the readiness of the model for detailed testing.

7. Documentation and Reporting

Objective: Keep detailed records of the build and testing processes.

Tasks:

- **Maintain Build Logs**: Document each step of the build process. Include any adjustments made during construction.
- **Update Design Documentation**: Ensure that any changes to the original design are documented, including new schematics or CAD drawings.
- **Prepare Progress Reports**: Submit progress reports to the professor or sponsor weekly to show the current status of the build.

Deliverable: Comprehensive documentation covering the entire build process, including updated designs, build logs, and testing results.

Timeline Overview:
- **Week 1**: Project planning and material procurement.
- **Week 2**: Begin assembly of subsystems.
- **Week 3**: Complete integration and initial testing.
- **Week 4**: Finalize the build and prepare for testing.

Final Submission Requirements:
- **Project Plan**: Gantt chart with tasks, responsibilities, and deadlines.
- **Build Plan**: Detailed step-by-step instructions for constructing the working model.
- **BOM**: Complete bill of materials and budget breakdown.
- **Risk Management Plan**: Document outlining potential risks and mitigation strategies.

- **Progress Reports**: Weekly updates during the build phase.
- **Final Testing Report**: Documented results of initial tests and validation.

This assignment will require effective teamwork, detailed planning, and a proactive approach to problem-solving. Meeting the 4-week deadline is essential for ensuring the working model is ready for test engineering activities at the start of February.

Assignment 15-2: Design Mistake-Proofing (Poka-Yoke) Features

Objective:

The purpose of this assignment is for your design team to critically analyze your prototype to identify potential user errors or misuse. Your task is to integrate **Poka-Yoke** (mistake-proofing) features to prevent these errors and improve the usability, safety, and reliability of the design. This approach ensures that your product performs as intended, even in the hands of non-expert users, and enhances user satisfaction.

Instructions:

Step 1: Analyze Your Design for Potential Misuse or Mistakes
- Review your prototype and identify areas where users may make mistakes (e.g., incorrect assembly, wrong input, improper use).
- Think about common challenges in similar products. Are there features that are unclear or could lead to operational errors?

Deliverable:
- Create a **list of potential user mistakes**. Include both minor mistakes (like misalignment) and major ones (such as safety risks).
- Briefly explain the impact of each mistake on product performance or user safety.

Step 2: Develop a Mistake-Proofing Strategy
- Select one or more of the identified mistakes to address with a Poka-Yoke solution.
- Brainstorm solutions that will **either prevent the mistake** or **make it obvious when the user makes an error**.
- Consider both **mechanical solutions** (e.g., shapes that fit only one way) and **software solutions** (e.g., alerts or error messages).

Deliverable:
- Document **three Poka-Yoke ideas**. Briefly describe how each feature would work and what mistake it would prevent.

Step 3: Choose the Best Poka-Yoke Solution
- Select the most practical and effective Poka-Yoke idea from Step 2.
- Justify your choice by considering factors such as cost, complexity, user experience, and manufacturability.

Deliverable:
- Write a **justification report** (200-300 words) explaining why this mistake-proofing feature is the most suitable for your design.

Step 4: Incorporate the Chosen Poka-Yoke into Your Design
- Make necessary modifications to your prototype or product design to integrate the selected Poka-Yoke feature.

- If the solution cannot be fully implemented in the prototype due to constraints, create a **detailed plan** for its integration in the final product.

Deliverable:
- Provide **updated design drawings** or sketches showing how the Poka-Yoke feature fits into the product.

Step 5: Test the New Design for Effectiveness
- Conduct a simulation or test with the modified design to ensure the Poka-Yoke feature works as intended.
- Gather feedback from users or team members to see if the error is successfully prevented.

Deliverable:
- Submit a **test report** summarizing:
    - The testing method used
    - Observations and results
    - Any remaining issues or recommendations for further refinement

Step 6: Document the Entire Process
- Create a brief report summarizing the entire Poka-Yoke development process, including identified mistakes, the chosen solution, and the results of the testing phase.

Deliverable:
- **Final report** (500-800 words) with images, sketches, or diagrams as necessary.

Evaluation Criteria:
- **Completeness**: Were all steps followed, and were deliverables submitted on time?
- **Depth of Analysis**: Did the team identify relevant user mistakes and explore practical Poka-Yoke solutions?
- **Creativity**: Are the proposed mistake-proofing features innovative and effective?
- **Implementation Feasibility**: Can the chosen Poka-Yoke feature be integrated into the final product with minimal issues?
- **Testing and Documentation**: Were the tests thorough, and was the process well-documented?

Resources:

For additional guidance, refer to **Section 15.6.7 of the workbook** which discusses Poka-Yoke in detail. Use examples from USB connectors or other products provided in the text to inspire your solutions.

16 Test Engineering

In the engineering design process, test planning, test engineering, and creating a testing matrix are pivotal activities that ensure the reliability, functionality, and safety of the final product. These activities are essential to validate design decisions, identify potential issues, and guide the product development process towards successful implementation.

- **Test Planning**

Test planning is the process of defining the scope, approach, resources, and schedule for testing activities. It involves the identification of key tests that need to be performed, the objectives of these tests, and the criteria for success. The test plan serves as a roadmap for the testing phase, outlining the sequence of tests and specifying the conditions under which they will be conducted.

Key components of a test plan include:
- **Test Objectives**: Clear statements of what each test aims to achieve.
- **Test Items**: Specific components or systems that will be tested.
- **Test Schedule**: A timeline for when each test will be conducted.
- **Resources**: Personnel, equipment, and materials required for testing.
- **Criteria for Success**: Metrics or standards that determine the success of the tests.

Effective test planning ensures that all necessary tests are identified and scheduled, resources are allocated appropriately, and potential risks are mitigated.

- **Test Engineering**

Test engineering involves the practical execution of the test plan. It encompasses the design, development, and implementation of tests to evaluate the performance, reliability, and safety of the product. Test engineers are responsible for setting up test environments, developing test procedures, and using various testing methods to gather data.

Some common testing methods include:
- **Prototype Testing**: Evaluating physical prototypes to verify design decisions.

- **Modeling and Simulations**: Using computational models to simulate real-world conditions.
- **Failure Mode Testing**: Testing components and systems to their limits to identify failure modes.
- **Environmental Testing**: Assessing the product's performance under extreme conditions such as high temperatures, humidity, or vibration.

Test engineering is critical for identifying design flaws and areas for improvement, ensuring that the product meets all specifications and requirements before it goes to market.

- **Creating a Testing Matrix**

A testing matrix, or test matrix, is a tool used to organize and manage the various tests that need to be conducted. It provides a structured way to ensure that all aspects of the product are tested thoroughly and systematically. The matrix typically includes the following elements:
   - **Test Cases**: Specific scenarios or conditions under which the product will be tested.
   - **Test Methods**: Procedures or techniques used to perform the tests.
   - **Criteria for Evaluation**: Standards or benchmarks used to assess test results.
   - **Responsibility**: Assignment of personnel responsible for conducting each test.

Creating a testing matrix involves:
1. **Defining Test Cases**: Based on the requirements and specifications of the product.
2. **Mapping Test Methods**: To each test case to ensure the appropriate techniques are used.
3. **Setting Evaluation Criteria**: To determine the pass/fail status of each test.
4. **Assigning Responsibilities**: To ensure accountability and clarity in the testing process.

The testing matrix helps in tracking the progress of testing activities, ensuring comprehensive coverage of all product aspects, and facilitating communication among team members.

In conclusion, test planning, test engineering, and creating a testing matrix are integral parts of the engineering design process that help in validating the product design, identifying potential issues, and ensuring that the final product meets all specified requirements and standards. These activities provide a systematic approach to testing, contributing to the development of reliable and high-quality products.

16.1 Test Planning

Test planning is a crucial phase in the engineering design process that ensures the reliability, safety, and functionality of the final product. It involves the systematic preparation for testing activities by defining objectives, identifying resources, scheduling tasks, and establishing success criteria. This section provides an in-depth exploration of test planning, highlighting its significance, methodologies, and best practices.

16.1.1 Importance of Test Planning

Test planning is essential for several reasons:

- **Risk Mitigation**: By identifying potential issues early in the design process, test planning helps mitigate risks that could lead to product failure.
- **Resource Optimization**: Proper planning ensures efficient use of resources, including time, personnel, and equipment.
- **Quality Assurance**: A well-structured test plan ensures that all product specifications and requirements are thoroughly tested, leading to higher quality products.
- **Cost Efficiency**: Identifying and addressing issues early can significantly reduce the costs associated with late-stage design changes and recalls.

16.1.2 Components of a Test Plan

A comprehensive test plan typically includes the following components:

1. **Test Objectives**: Clear and concise statements outlining what each test aims to achieve. Objectives should be specific, measurable, achievable, relevant, and time-bound (SMART).
2. **Test Scope**: Defines the boundaries of the testing activities, specifying what will and will not be tested. This helps in focusing efforts on critical areas and managing stakeholder expectations.
3. **Test Items**: A detailed list of components, subsystems, or systems that will be tested. This includes information on the version or configuration of each item to be tested.
4. **Test Schedule**: A timeline outlining when each test will be conducted. It includes start and end dates, dependencies between tests, and key milestones.
5. **Resources**: Details of personnel, equipment, materials, and facilities required for testing. This ensures that all necessary resources are available when needed.
6. **Criteria for Success**: Metrics or standards used to evaluate the test results. These criteria help in determining whether a test has passed or failed.
7. **Risk Management**: Identification of potential risks that could impact the testing activities and strategies for mitigating these risks.

8. **Documentation and Reporting**: Specifies the format and content of test reports, including how results will be documented and communicated to stakeholders.

16.1.3 Methodologies in Test Planning

Test planning involves various methodologies to ensure comprehensive coverage and effectiveness:

1. **Requirements-Based Testing**: This methodology focuses on deriving test cases directly from the product requirements. It ensures that all specified functionalities and performance criteria are validated.
2. **Risk-Based Testing**: Prioritizes testing activities based on the risk associated with different components or functionalities. Higher-risk areas are tested more rigorously to minimize the chances of critical failures.
3. **Boundary Value Analysis**: Involves testing the product at the extreme values of input parameters. This helps in identifying issues that occur at the boundaries of operational limits.
4. **Equivalence Partitioning**: Divides input data into equivalent partitions that are expected to behave similarly. Tests are conducted on representative values from each partition, reducing the number of test cases while maintaining coverage.
5. **State Transition Testing**: Used for systems with distinct states and transitions. Test cases are designed to cover all possible state transitions, ensuring that the system behaves correctly in each state.
6. **Exploratory Testing**: An informal approach where testers explore the product functionality without predefined test cases. It helps in identifying unexpected issues and gaining a deeper understanding of the product.

16.1.4 Best Practices in Test Planning

To ensure the effectiveness of test planning, the following best practices should be adhered to:

1. **Early Involvement**: Involve test planning early in the design process. Early involvement helps in identifying potential issues sooner and integrating testing activities seamlessly into the development process.
2. **Stakeholder Collaboration**: Collaborate with stakeholders, including designers, developers, and end-users, to understand their expectations and incorporate their input into the test plan.
3. **Clear Documentation**: Maintain clear and detailed documentation of the test plan. This serves as a reference for the testing team and ensures consistency in testing activities.
4. **Continuous Review and Update**: Regularly review and update the test plan

to reflect changes in design, requirements, or testing conditions. This ensures that the test plan remains relevant and effective.
5. **Automation**: Where feasible, incorporate automated testing tools and frameworks. Automation can significantly enhance the efficiency and coverage of testing activities.
6. **Metrics and Monitoring**: Define metrics to monitor the progress and effectiveness of testing activities. Regular monitoring helps in identifying bottlenecks and areas for improvement.
7. **Training and Skill Development**: Ensure that the testing team is well-trained and equipped with the necessary skills and knowledge. Continuous skill development helps in keeping up with evolving testing methodologies and tools.

16.1.5 Example of a Test Plan

The following example illustrates a test plan for a hypothetical mechanical component design project.

Project Name: Gearbox Design

Test Objectives:
1. Validate the structural integrity of the gearbox under different load conditions.
2. Ensure the gearbox operates smoothly without excessive noise or vibration.
3. Verify the lubrication system maintains adequate lubrication under all operating conditions.

Test Scope:
- Included: Structural analysis, noise and vibration testing, lubrication system testing.
- Excluded: Aesthetic evaluations, packaging tests.

Test Items:
1. Gearbox assembly (Version 1.0)
2. Lubrication system components

Test Schedule:

Test Activity	Start Date	End Date	Dependencies
Structural Analysis	01/08/2024	15/08/2024	None
Noise and Vibration Testing	16/08/2024	25/08/2024	Structural Analysis
Lubrication System Testing	26/08/2024	05/09/2024	Noise and Vibration

Resources:
- Personnel: 2 Test Engineers, 1 Mechanical Engineer
- Equipment: Universal testing machine, accelerometer, lubrication test rig
- Materials: Gearbox assemblies, lubrication oil

Criteria for Success:
1. Structural Analysis: No deformation or failure under maximum load.
2. Noise and Vibration: Noise level below 75 dB, vibration amplitude within specified limits.
3. Lubrication System: Continuous lubrication without interruption for 24 hours.

Risk Management:
- Potential Delays: Mitigated by early procurement of testing equipment and materials.
- Equipment Failure: Backup equipment available and regular maintenance scheduled.

Documentation and Reporting:
- Test reports to include detailed test procedures, results, and observations.
- Weekly progress reports to be shared with the project team and stakeholders.

Test planning is an integral part of the engineering design process that ensures the final product meets its intended performance, reliability, and safety standards. By systematically defining objectives, resources, schedules, and success criteria, test planning provides a structured approach to validate design decisions and identify potential issues early in the development process. Adopting best practices and methodologies in test planning enhances the efficiency and effectiveness of testing activities, ultimately leading to the delivery of high-quality products.

16.2 Test Plan Development

In engineering capstone design projects, test plan development is a critical phase that ensures the final product meets its intended performance, reliability, and safety requirements. A well-structured test plan helps in systematically verifying and validating the design, identifying potential issues early, and ensuring that the project stays on track. This section provides an in-depth exploration of test plan development, focusing on its significance, methodologies, components, and best practices.

16.2.1 Importance of Test Plan Development

Developing a test plan is essential in capstone design projects for several reasons:
- **Validation of Design**: Ensures that the design meets the specified requirements and functions as intended.
- **Early Detection of Issues**: Identifies potential design flaws and performance issues early in the project, allowing for timely corrections.
- **Resource Management**: Helps in the efficient allocation and utilization of resources, including time, personnel, and equipment.
- **Documentation and Communication**: Provides a clear and structured documentation of the testing process, facilitating better communication among

16.2 Test Plan Development

team members and stakeholders.
- **Risk Mitigation**: Reduces the risk of project failure by systematically testing and validating each aspect of the design.

16.2.2 Components of a Test Plan

A comprehensive test plan for a capstone design project typically includes the following components:

1. **Test Objectives**:
    - Define the specific goals of the testing activities.
    - Ensure that all critical aspects of the design are tested.
    - Examples: Verify the structural integrity, ensure compliance with safety standards, validate functional performance.
2. **Test Scope**:
    - Clearly delineate the boundaries of the testing activities.
    - Specify what will and will not be tested.
    - This helps in focusing efforts on critical areas and managing expectations.
3. **Test Items**:
    - List the specific components, subsystems, or systems that will be tested.
    - Include details such as version numbers, configurations, and any relevant documentation.
4. **Test Environment**:
    - Describe the environment in which the tests will be conducted.
    - Include details about the physical setup, software environment, and any necessary conditions for testing.
5. **Test Schedule**:
    - Provide a timeline for the testing activities.
    - Include start and end dates, dependencies between tests, and key milestones.
    - Ensure that the schedule is realistic and allows for adequate time for each testing activity.
6. **Resources**:
    - Detail the personnel, equipment, materials, and facilities required for testing.
    - Ensure that all necessary resources are available and allocated appropriately.
7. **Test Procedures**:
    - Describe the step-by-step procedures for conducting each test.
    - Include details about the setup, execution, and teardown of each test.

o Ensure that the procedures are clear and repeatable.
8. **Criteria for Success**:
    o Define the metrics or standards used to evaluate the test results.
    o Ensure that the criteria are specific, measurable, and aligned with the project goals.
9. **Risk Management**:
    o Identify potential risks that could impact the testing activities.
    o Develop strategies for mitigating these risks.
10. **Documentation and Reporting**:
    o Specify the format and content of test reports.
    o Include details about how results will be documented and communicated to stakeholders.

16.2.3 Methodologies in Test Plan Development

Various methodologies can be employed in test plan development to ensure comprehensive coverage and effectiveness:

1. **Requirements-Based Testing**:
    o Derive test cases directly from the project requirements.
    o Ensure that all specified functionalities and performance criteria are validated.
2. **Risk-Based Testing**:
    o Prioritize testing activities based on the risk associated with different components or functionalities.
    o Focus more rigorous testing on higher-risk areas to minimize the chances of critical failures.
3. **Boundary Value Analysis**:
    o Test the design at the extreme values of input parameters.
    o Identify issues that occur at the boundaries of operational limits.
4. **Equivalence Partitioning**:
    o Divide input data into equivalent partitions that are expected to behave similarly.
    o Test representative values from each partition to reduce the number of test cases while maintaining coverage.
5. **State Transition Testing**:
    o Use for systems with distinct states and transitions.
    o Design test cases to cover all possible state transitions, ensuring correct system behavior in each state.
6. **Exploratory Testing**:
    o An informal approach where testers explore the design functionality without predefined test cases.

- Helps in identifying unexpected issues and gaining a deeper understanding of the design.

16.2.4 Best Practices in Test Plan Development

To ensure the effectiveness of test plan development in capstone design projects, the following best practices should be adhered to:

1. **Early Involvement**:
   - Involve test planning early in the design process.
   - Early involvement helps in identifying potential issues sooner and integrating testing activities seamlessly into the development process.
2. **Stakeholder Collaboration**:
   - Collaborate with all stakeholders, including advisors, team members, and end-users, to understand their expectations and incorporate their input into the test plan.
3. **Clear Documentation**:
   - Maintain clear and detailed documentation of the test plan.
   - This serves as a reference for the testing team and ensures consistency in testing activities.
4. **Continuous Review and Update**:
   - Regularly review and update the test plan to reflect changes in design, requirements, or testing conditions.
   - This ensures that the test plan remains relevant and effective.
5. **Automation**:
   - Where feasible, incorporate automated testing tools and frameworks.
   - Automation can significantly enhance the efficiency and coverage of testing activities.
6. **Metrics and Monitoring**:
   - Define metrics to monitor the progress and effectiveness of testing activities.
   - Regular monitoring helps in identifying bottlenecks and areas for improvement.
7. **Training and Skill Development**:
   - Ensure that the testing team is well-trained and equipped with the necessary skills and knowledge.
   - Continuous skill development helps in keeping up with evolving testing methodologies and tools.

16.2.5 Example of a Test Plan

The following example illustrates a test plan for a hypothetical capstone design project involving the development of a new type of drone:

Project Name: Autonomous Delivery Drone

Test Objectives:
1. Validate the flight stability and control under various weather conditions.
2. Ensure the payload delivery mechanism functions reliably.
3. Verify the battery life and range meet the specified requirements.

Test Scope:
- Included: Flight stability, control system, payload delivery mechanism, battery performance.
- Excluded: Aesthetic design evaluations, user interface testing.

Test Items:
1. Drone prototype (Version 1.0)
2. Control system software
3. Payload delivery mechanism

Test Environment:
- Outdoor testing area with variable weather conditions.
- Indoor testing facility for controlled environment tests.
- Simulation software for initial validation of control algorithms.

Test Schedule:

Test Activity	Start Date	End Date	Dependencies
Control System Testing	01/09/2024	10/09/2024	None
Flight Stability Testing	11/09/2024	20/09/2024	Control System Testing
Payload Delivery Mechanism Testing	21/09/2024	30/09/2024	Flight Stability Testing
Battery Performance Testing	01/10/2024	10/10/2024	Payload Delivery

Resources:
- Personnel: 3 Test Engineers, 1 Software Engineer
- Equipment: Drone prototypes, weather simulation chamber, battery test rig
- Materials: Batteries, payload items

Test Procedures:
1. **Control System Testing**:
   - Set up the control system in the simulation software.
   - Run a series of pre-defined maneuvers to validate control algorithms.
   - Document any deviations and refine the control system.
2. **Flight Stability Testing**:
   - Conduct test flights in the outdoor testing area under various weather conditions.
   - Measure stability metrics such as yaw, pitch, and roll.
   - Compare results against specified thresholds and make

16.2 Test Plan Development

            necessary adjustments.
3. **Payload Delivery Mechanism Testing**:
   - Test the payload delivery mechanism in both controlled and real-world environments.
   - Ensure the mechanism operates reliably and securely releases the payload.
   - Record any failures and implement design improvements.
4. **Battery Performance Testing**:
   - Measure the battery life under different flight conditions and payload weights.
   - Ensure the battery meets the specified range and duration.
   - Analyze any discrepancies and identify potential enhancements.

Criteria for Success:
1. Control System Testing: Control algorithms function correctly in the simulation.
2. Flight Stability Testing: Stability metrics remain within acceptable limits under all tested conditions.
3. Payload Delivery Mechanism Testing: Payload is delivered reliably without failures.
4. Battery Performance Testing: Battery life and range meet or exceed specified requirements.

Risk Management:
- **Adverse Weather Conditions**: Mitigate by scheduling outdoor tests in advance and having backup indoor tests.
- **Equipment Failure**: Ensure backup equipment is available and perform regular maintenance.

Documentation and Reporting:
- **Test Reports**: Include detailed test procedures, results, and observations for each test.
- **Progress Reports**: Share weekly updates with the project team and stakeholders, highlighting any issues and corrective actions.

Test plan development is a critical component of engineering capstone design projects, ensuring that the final product meets its intended performance, reliability, and safety requirements. By systematically defining objectives, resources, schedules, and success criteria, test planning provides a structured approach to validate design decisions and identify potential issues early in the development process. Adopting best practices and methodologies in test planning enhances the efficiency and effectiveness of testing activities, ultimately leading to the successful completion of the capstone design project.

16.3 Test Engineering Matrix

A test engineering matrix is a crucial tool in the capstone design project process. It organizes and structures the various tests required to validate a product's design, ensuring all aspects of the product are thoroughly evaluated. This section delves into the creation, implementation, and management of a test engineering matrix, emphasizing its significance in engineering design, the methodologies involved, and best practices for its development and use.

16.3.1 Importance of a Test Engineering Matrix

A test engineering matrix provides several key benefits:
- **Comprehensive Coverage**: Ensures all critical components and functionalities are tested.
- **Structured Testing**: Organizes tests logically, helping to track progress and manage testing activities efficiently.
- **Clear Documentation**: Provides a clear record of testing procedures, results, and responsibilities.
- **Resource Allocation**: Helps in planning and allocating resources effectively.
- **Risk Mitigation**: Identifies potential issues early, allowing for timely corrective actions.

16.3.2 Components of a Test Engineering Matrix

A comprehensive test engineering matrix includes the following components:
1. **Test Cases**:
    - Specific scenarios or conditions under which the product will be tested.
    - Detailed descriptions of each test, including inputs, expected outcomes, and criteria for evaluation.
2. **Test Methods**:
    - Procedures or techniques used to conduct each test.
    - Includes information on equipment, tools, and methodologies to be used.
3. **Criteria for Evaluation**:
    - Standards or benchmarks used to assess the results of each test.
    - Defines what constitutes a pass or fail for each test case.
4. **Test Environment**:
    - Description of the environment in which tests will be conducted.
    - Includes physical setup, software environment, and any special

conditions required for testing.
5. **Responsibilities**:
    o Assignment of personnel responsible for conducting each test.
    o Ensures accountability and clarity in the testing process.
6. **Schedule**:
    o Timeline for conducting each test.
    o Includes start and end dates, dependencies, and milestones.
7. **Results and Documentation**:
    o Detailed recording of test results.
    o Documentation of any deviations, issues, and corrective actions taken.

16.3.3 Developing a Test Engineering Matrix

Developing a test engineering matrix involves several steps:
1. **Identify Test Cases**:
    o Begin by identifying all the key components and functionalities of the product.
    o For each component or functionality, define specific test cases that need to be conducted.
    o Ensure that the test cases cover normal, boundary, and failure scenarios.
2. **Define Test Methods**:
    o For each test case, define the method or procedure to be used.
    o Include details about the equipment, tools, and techniques required for testing.
    o Ensure that the test methods are practical, repeatable, and aligned with industry standards.
3. **Set Criteria for Evaluation**:
    o Define clear criteria for evaluating the results of each test.
    o Criteria should be specific, measurable, and aligned with the project goals.
    o Include both quantitative metrics (e.g., performance thresholds) and qualitative criteria (e.g., user satisfaction).
4. **Establish Test Environment**:
    o Describe the environment in which each test will be conducted.
    o Ensure that the environment is suitable for the tests and that all necessary equipment and tools are available.
    o Include any special conditions or setups required for the tests.
5. **Assign Responsibilities**:
    o Assign personnel responsible for conducting each test.
    o Ensure that each team member understands their responsibilities and

has the necessary skills and knowledge.
- Include contact information and any relevant qualifications or experience.
6. **Create Schedule**:
   - Develop a timeline for conducting each test.
   - Include start and end dates, dependencies between tests, and key milestones.
   - Ensure that the schedule is realistic and allows for adequate time for each testing activity.
7. **Document Results**:
   - Develop a format for recording and documenting test results.
   - Include sections for test procedures, results, deviations, issues, and corrective actions.
   - Ensure that the documentation is clear, detailed, and easy to follow.

16.3.4 Example of a Test Engineering Matrix

The following example illustrates a test engineering matrix for a capstone design project involving the development of an autonomous delivery drone.

Project Name: Autonomous Delivery Drone

16.3 Test Engineering Matrix

Test Engineering Matrix:

Test Case	Test Method	Criteria for Evaluation	Test Environment	Responsibilities	Schedule	Results and Documentation
Flight Stability Under Various Conditions	Conduct test flights in outdoor testing area under different weather conditions	Stability metrics (yaw, pitch, roll) within specified thresholds	Outdoor testing area, variable weather conditions	Test Engineer A	01/09/2024 - 05/09/2024	Record stability metrics for each flight, document any deviations, and corrective actions taken
Payload Delivery Mechanism Reliability	Test the payload delivery mechanism in controlled and real-world environments	Mechanism operates reliably, payload securely released	Indoor facility and outdoor testing area	Test Engineer B	06/09/2024 - 10/09/2024	Document number of successful deliveries, record any failures and corresponding corrective actions
Battery Life and Range	Measure battery performance under different flight conditions and payload weights	Battery life and range meet or exceed specified requirements	Indoor testing facility, battery test rig	Test Engineer C	11/09/2024 - 15/09/2024	Record battery performance metrics, document any discrepancies, and potential enhancements
Control System Software Validation	Run control system in simulation software, validate control algorithms	Control algorithms function correctly in the simulation	Simulation software	Software Engineer	16/09/2024 - 20/09/2024	Document control system performance in simulation, note any issues, and adjustments made
Environmental Impact Assessment	Evaluate drone's impact on the environment in terms of noise, emissions, and wildlife interaction	Environmental impact within acceptable limits	Outdoor testing area, various natural settings	Environmental Engineer	21/09/2024 - 25/09/2024	Document environmental impact findings, provide recommendations for minimizing impact
Emergency Landing Protocols	Test emergency landing procedures under different failure scenarios	Drone lands safely and minimizes damage	Outdoor testing area, simulation software	Safety Engineer	26/09/2024 - 30/09/2024	Record performance of emergency landing protocols, document any issues and corrective measures
User Interface Usability	Conduct usability tests with a group of representative users	User satisfaction ratings, ease of use	Indoor testing facility	Human Factors Engineer	01/10/2024 - 05/10/2024	Collect user feedback, record usability issues, and suggest improvements

8. **Detailed Explanation of Test Cases**
    1. **Flight Stability Under Various Conditions**:
        - **Objective**: Validate the drone's flight stability under different weather conditions.
        - **Method**: Conduct test flights in an outdoor testing area with varying wind speeds, temperatures, and humidity levels. Use sensors to measure yaw, pitch, and roll.
        - **Criteria**: Stability metrics should remain within specified thresholds under all tested conditions.
        - **Responsibilities**: Test Engineer A is responsible for setting up the test environment, conducting the flights, and recording the data.
    2. **Payload Delivery Mechanism Reliability**:
        - **Objective**: Ensure the payload delivery mechanism functions reliably in both controlled and real-world environments.
        - **Method**: Test the mechanism in an indoor facility with controlled conditions, then conduct tests in outdoor settings. Monitor the mechanism's performance during multiple delivery attempts.
        - **Criteria**: The mechanism should operate reliably and securely release the payload in at least 95% of attempts.
        - **Responsibilities**: Test Engineer B will oversee the testing process, document the number of successful deliveries, and record any failures.
    3. **Battery Life and Range**:
        - **Objective**: Verify that the battery life and range meet the specified requirements under different flight conditions and payload weights.
        - **Method**: Conduct tests in an indoor testing facility using a battery test rig. Measure the battery performance during different flight scenarios and with varying payload weights.
        - **Criteria**: The battery life should be sufficient for a minimum of 30 minutes of flight, and the range should meet the specified requirements under all tested conditions.
        - **Responsibilities**: Test Engineer C will conduct the tests, record the battery performance metrics, and document any discrepancies.
    4. **Control System Software Validation**:
        - **Objective**: Validate the control system software using simulation software and pre-defined maneuvers.
        - **Method**: Run the control system in the simulation software and perform a series of maneuvers to validate the control algorithms. Analyze the system's response to each maneuver.
        - **Criteria**: The control algorithms should function correctly in the

simulation, with no significant deviations from the expected behavior.
- **Responsibilities**: The Software Engineer will be responsible for setting up the simulation, running the tests, and documenting the system's performance.

5. **Environmental Impact Assessment**:
   - **Objective**: Assess the drone's impact on the environment in terms of noise, emissions, and wildlife interaction.
   - **Method**: Conduct tests in various natural settings to measure the noise levels, emissions, and interactions with wildlife. Use sound level meters and environmental monitoring equipment.
   - **Criteria**: The environmental impact should be within acceptable limits, as defined by regulatory standards.
   - **Responsibilities**: The Environmental Engineer will conduct the assessments, document the findings, and provide recommendations for minimizing the environmental impact.

6. **Emergency Landing Protocols**:
   - **Objective**: Test the emergency landing procedures under different failure scenarios to ensure the drone can land safely and minimize damage.
   - **Method**: Simulate various failure scenarios, such as loss of control, power failure, and communication loss. Monitor the drone's response and landing performance.
   - **Criteria**: The drone should be able to land safely and minimize damage in at least 90% of the simulated scenarios.
   - **Responsibilities**: The Safety Engineer will set up the simulations, conduct the tests, and document the performance of the emergency landing protocols.

7. **User Interface Usability**:
   - **Objective**: Evaluate the usability of the user interface by conducting tests with a group of representative users.
   - **Method**: Conduct usability tests in an indoor testing facility, where users perform a series of tasks using the drone's interface. Collect feedback on the ease of use and user satisfaction.
   - **Criteria**: User satisfaction ratings should be above 80%, and the interface should be rated as easy to use by at least 90% of the users.
   - **Responsibilities**: The Human Factors Engineer will conduct the usability tests, collect user feedback, and document any usability issues.

16.3.5 Best Practices for Using a Test Engineering Matrix

To maximize the effectiveness of a test engineering matrix, the following best practices should be adopted:

1. **Regular Updates**:
   o Keep the matrix updated throughout the project to reflect changes in design, requirements, or testing conditions.
   o Regular updates ensure that the matrix remains relevant and useful.
2. **Clear Communication**:
   o Ensure that all team members understand the matrix and their responsibilities.
   o Use the matrix as a tool for facilitating communication and coordination among team members.
3. **Comprehensive Documentation**:
   o Document all test procedures, results, deviations, and corrective actions in detail.
   o Comprehensive documentation helps in tracking progress, identifying issues, and making informed decisions.
4. **Continuous Review and Improvement**:
   o Regularly review the matrix to identify areas for improvement.
   o Use feedback from testing activities to refine and enhance the matrix.
5. **Integration with Project Management Tools**:
   o Integrate the test engineering matrix with project management tools to streamline tracking and reporting.
   o Integration helps in aligning testing activities with the overall project schedule and milestones.
6. **Use of Automation**:
   o Incorporate automated testing tools and frameworks where feasible.
   o Automation can enhance the efficiency and coverage of testing activities, reducing manual effort and minimizing errors.
7. **Focus on Risk Management**:
   o Use the matrix to identify and manage risks associated with testing activities.
   o Develop and implement risk mitigation strategies to address potential issues proactively.

A test engineering matrix is a vital tool in capstone design projects, providing a structured and organized approach to testing. By detailing test cases, methods, criteria, environment, responsibilities, schedule, and documentation, the matrix ensures comprehensive coverage and effective management of testing activities. Adopting best practices and methodologies in the development and use of a test engineering matrix enhances the quality and reliability of the final product, contributing to the successful completion of the capstone design project.

16.4 Test the Design and the Build

Testing the design and build of a prototype in a capstone mechanical engineering project is a multifaceted process that involves verifying that the prototype meets design specifications and functions as intended under various conditions. This phase is crucial for validating the design, identifying potential issues, and ensuring that the final product is reliable, efficient, and safe. This section provides an in-depth exploration of the methodologies, steps, and best practices involved in testing the design and build of a prototype in mechanical engineering.

16.4.1 Importance of Prototype Testing

Prototype testing in mechanical engineering serves several essential purposes:
- **Validation**: Confirms that the prototype meets design specifications and performs as intended.
- **Verification**: Ensures that the prototype functions correctly under various conditions and scenarios.
- **Risk Reduction**: Identifies potential issues early in the development process, allowing for timely corrections.
- **Optimization**: Provides insights into areas for improvement, enhancing the overall design.
- **Compliance**: Ensures that the prototype complies with relevant standards and regulations.

16.4.2 Methodologies in Prototype Testing

Various methodologies are employed to test the design and build of a prototype comprehensively:
1. **Functional Testing**:
    - **Objective**: Verify that the prototype performs its intended functions correctly.
    - **Method**: Test each function of the prototype under normal and extreme conditions.
    - **Criteria**: The prototype should meet all functional requirements and perform reliably under all tested conditions.
2. **Performance Testing**:
    - **Objective**: Evaluate the performance characteristics of the prototype,

such as speed, efficiency, and power consumption.
- o **Method**: Measure the prototype's performance metrics under various operating conditions.
- o **Criteria**: The prototype should meet or exceed the specified performance benchmarks.

3. **Stress Testing**:
   - o **Objective**: Assess the prototype's ability to withstand extreme conditions and stress.
   - o **Method**: Subject the prototype to conditions beyond its normal operating limits, such as high loads, extreme temperatures, and prolonged operation.
   - o **Criteria**: The prototype should function correctly without failure or significant degradation under extreme conditions.

4. **Durability Testing**:
   - o **Objective**: Determine the longevity and wear resistance of the prototype.
   - o **Method**: Conduct long-term tests to evaluate the prototype's durability and identify potential points of failure.
   - o **Criteria**: The prototype should maintain its performance and integrity over the expected lifespan.

5. **Usability Testing**:
   - o **Objective**: Evaluate the ease of use and user satisfaction with the prototype.
   - o **Method**: Conduct tests with representative users to gather feedback on usability and identify any issues.
   - o **Criteria**: The prototype should be user-friendly and meet user expectations for functionality and ease of use.

6. **Environmental Testing**:
   - o **Objective**: Assess the prototype's performance under different environmental conditions.
   - o **Method**: Test the prototype in various environments, such as high humidity, dust, and vibration.
   - o **Criteria**: The prototype should perform reliably under all tested environmental conditions.

7. **Compliance Testing**:
   - o **Objective**: Ensure the prototype complies with relevant standards and regulations.
   - o **Method**: Conduct tests to verify compliance with industry standards, safety regulations, and other requirements.
   - o **Criteria**: The prototype should meet all applicable standards and

16.4 Test the Design and the Build

regulatory requirements.

16.4.3 Steps in Testing the Prototype

Testing the prototype involves several key steps, each of which ensures a thorough and systematic evaluation of the design and build:

1. **Test Planning**:
   o Develop a detailed test plan outlining the objectives, scope, methods, and criteria for each test.
   o Identify the resources required for testing, including personnel, equipment, and materials.
   o Establish a timeline for conducting the tests, including start and end dates, dependencies, and milestones.
2. **Setup and Preparation**:
   o Prepare the test environment, ensuring that all necessary equipment and tools are available and in working order.
   o Assemble the prototype and verify that it is ready for testing.
   o Ensure that all personnel involved in testing are trained and familiar with the test procedures.
3. **Conducting Tests**:
   o Perform each test according to the test plan, following the specified procedures and methods.
   o Monitor the prototype's performance and record the results of each test.
   o Identify any deviations from the expected performance and document any issues or failures.
4. **Data Collection and Analysis**:
   o Collect data from each test and analyze the results to determine the prototype's performance.
   o Compare the test results with the specified criteria and benchmarks.
   o Identify any trends or patterns in the data that may indicate potential issues or areas for improvement.
5. **Documentation and Reporting**:
   o Document the test procedures, results, and any issues or failures in detail.
   o Prepare a comprehensive test report summarizing the findings and providing recommendations for improvements.
   o Share the test report with stakeholders, including the project team, advisors, and any relevant regulatory bodies.
6. **Review and Iteration**:
   o Review the test results and identify any necessary design changes or

improvements.
- o Implement the recommended changes and prepare a revised prototype.
- o Repeat the testing process with the revised prototype to verify the effectiveness of the changes.

16.4.4 Detailed Example of Prototype Testing

The following example illustrates a detailed test plan for a hypothetical capstone design project involving the development of a robotic arm for precision manufacturing.

Project Name: Precision Manufacturing Robotic Arm

Test Plan:

Test Objective: Validate the design and performance of the robotic arm prototype.

Test Scope: Functional testing, performance testing, stress testing, durability testing, usability testing, environmental testing, and compliance testing.

Test Cases:

1. **Functional Testing**:
   - o **Test Case 1**: Validate the arm's precision in positioning.
     - **Method**: Use a coordinate measuring machine (CMM) to measure the arm's accuracy in reaching specified coordinates.
     - **Criteria**: The arm should position within ±0.01 mm of the target coordinates.
   - o **Test Case 2**: Validate the arm's repeatability.
     - **Method**: Perform multiple cycles of the same movement and measure the deviation.
     - **Criteria**: The standard deviation of the arm's endpoint should be less than 0.005 mm.
2. **Performance Testing**:
   - o **Test Case 3**: Measure the arm's speed.
     - **Method**: Use high-speed cameras to track the arm's movement and calculate the speed.
     - **Criteria**: The arm should achieve a speed of at least 1 m/s.
   - o **Test Case 4**: Evaluate the arm's power consumption.
     - **Method**: Measure the power draw during various operating conditions.
     - **Criteria**: The power consumption should be less than 500 W during normal operation.
3. **Stress Testing**:
   - o **Test Case 5**: Assess the arm's performance under high load.

- **Method**: Apply weights to the arm and monitor its performance.
- **Criteria**: The arm should lift weights up to 10 kg without significant deviation or damage.
- **Test Case 6**: Evaluate the arm's performance under prolonged operation.
  - **Method**: Run the arm continuously for 24 hours and monitor its performance.
  - **Criteria**: The arm should maintain its performance and functionality throughout the test period.

4. **Durability Testing**:
   - **Test Case 7**: Determine the longevity of the arm's joints and actuators.
     - **Method**: Perform repeated cycles of movement and monitor the wear on joints and actuators.
     - **Criteria**: The joints and actuators should function correctly without significant wear for at least 1 million cycles.
   - **Test Case 8**: Assess the wear resistance of the arm's surface.
     - **Method**: Subject the arm's surface to abrasive materials and measure the wear.
     - **Criteria**: The surface wear should be less than 0.1 mm after exposure to abrasive materials.

5. **Usability Testing**:
   - **Test Case 9**: Evaluate the ease of programming the arm.
     - **Method**: Have a group of users program the arm to perform specific tasks and gather feedback.
     - **Criteria**: User satisfaction ratings should be above 80%, and the programming process should be rated as easy to use by at least 90% of users.
   - **Test Case 10**: Assess the ease of maintenance.
     - **Method**: Conduct maintenance tasks on the arm and gather feedback on the ease of these tasks.
     - **Criteria**: Maintenance tasks should be completed within the expected time and with minimal difficulty.

6. **Environmental Testing**:
   - **Test Case 11**: Assess the arm's performance in different temperature conditions.
     - **Method**: Test the arm in a temperature-controlled chamber, varying temperatures from -20°C to 50°C.

- **Criteria**: The arm should perform reliably and maintain its precision across the tested temperature range.
- **Test Case 12**: Evaluate the arm's resistance to dust and humidity.
  - **Method**: Expose the arm to high humidity and dust environments and monitor its performance.
  - **Criteria**: The arm should perform reliably without significant degradation in high humidity and dust conditions.

7. **Compliance Testing**:
   - **Test Case 13**: Ensure the arm complies with safety regulations.
     - **Method**: Conduct tests to verify compliance with safety standards, such as ISO 10218 for industrial robots.
     - **Criteria**: The arm should meet all safety requirements specified in the standards.
   - **Test Case 14**: Verify the arm's electromagnetic compatibility (EMC).
     - **Method**: Test the arm for electromagnetic emissions and susceptibility.
     - **Criteria**: The arm should comply with EMC regulations, such as those specified in EN 61000-6-2 and EN 61000-6-4.

Steps in Conducting the Tests:
1. **Test Planning**:
   - Develop a detailed test plan outlining the objectives, scope, methods, and criteria for each test.
   - Identify the resources required for testing, including personnel, equipment, and materials.
   - Establish a timeline for conducting the tests, including start and end dates, dependencies, and milestones.
2. **Setup and Preparation**:
   - Prepare the test environment, ensuring that all necessary equipment and tools are available and in working order.
   - Assemble the prototype and verify that it is ready for testing.
   - Ensure that all personnel involved in testing are trained and familiar with the test procedures.
3. **Conducting Tests**:
   - Perform each test according to the test plan, following the specified procedures and methods.
   - Monitor the prototype's performance and record the results of

16.4 Test the Design and the Build

each test.
- o Identify any deviations from the expected performance and document any issues or failures.

4. **Data Collection and Analysis**:
   - o Collect data from each test and analyze the results to determine the prototype's performance.
   - o Compare the test results with the specified criteria and benchmarks.
   - o Identify any trends or patterns in the data that may indicate potential issues or areas for improvement.
5. **Documentation and Reporting**:
   - o Document the test procedures, results, and any issues or failures in detail.
   - o Prepare a comprehensive test report summarizing the findings and providing recommendations for improvements.
   - o Share the test report with stakeholders, including the project team, advisors, and any relevant regulatory bodies.
6. **Review and Iteration**:
   - o Review the test results and identify any necessary design changes or improvements.
   - o Implement the recommended changes and prepare a revised prototype.
   - o Repeat the testing process with the revised prototype to verify the effectiveness of the changes.

16.4.5 Best Practices in Prototype Testing

To ensure the effectiveness of prototype testing, the following best practices should be adhered to:

1. **Thorough Test Planning**:
   - o Develop a comprehensive test plan that covers all critical aspects of the prototype's performance.
   - o Ensure that the test plan is reviewed and approved by all relevant stakeholders.
2. **Clear Documentation**:
   - o Maintain clear and detailed documentation of all test procedures, results, and issues.
   - o Use standardized templates and formats for consistency and ease of use.
3. **Regular Updates and Reviews**:
   - o Regularly review and update the test plan and procedures to

reflect changes in the prototype design or testing conditions.
- o Conduct periodic reviews of the test results to identify any emerging issues or trends.
4. **Collaboration and Communication**:
    - o Foster collaboration and communication among all team members involved in testing.
    - o Ensure that any issues or concerns are promptly addressed and resolved.
5. **Use of Automation**:
    - o Where feasible, incorporate automated testing tools and frameworks to enhance efficiency and coverage.
    - o Automation can help reduce manual effort and minimize the risk of human error.
6. **Risk Management**:
    - o Identify and manage potential risks associated with testing activities.
    - o Develop and implement risk mitigation strategies to address potential issues proactively.
7. **Continuous Improvement**:
    - o Use the insights gained from testing to continuously improve the prototype design and testing processes.
    - o Encourage a culture of learning and innovation within the project team.

Testing the design and build of a prototype in a capstone mechanical engineering project is a comprehensive process that involves various methodologies and detailed steps to ensure the prototype meets the intended specifications and functions as required. By following best practices and systematically planning, conducting, and documenting tests, engineers can validate their designs, identify potential issues, and ensure the final product is reliable, efficient, and compliant with relevant standards and regulations. This rigorous approach to prototype testing is essential for the successful completion of capstone design projects and the development of high-quality mechanical engineering solutions.

16.6 Assignments

Assignment 16-1: Creating a Test Engineering Matrix and Running Statistical Tests

Objective:

This assignment requires your design team to develop a comprehensive **Test Engineering Matrix** and conduct detailed testing of key design parameters. You will execute each test multiple times to ensure consistency, identify variability, and perform statistical analysis on the results. Additionally, you will create a **Test Instance Recording Sheet** to document each test run. The goal is to complete your testing process within **2-4 weeks**.

Assignment Instructions

Part 1: Creating the Test Engineering Matrix

Step 1: Identify the Key Parameters to Test
- Review your design specifications, customer requirements, and problem definitions to identify **critical parameters** that need to be validated.
  Examples: Force, temperature, voltage, speed, load-bearing capacity, etc.
- Ensure the parameters selected are measurable and align with the performance goals of your prototype or product.

Deliverable:
- A **list of key parameters** to be tested and why they are essential to the product's success.

Step 2: Develop Test Scenarios
- For each parameter, create test scenarios that reflect real-world conditions where the product will be used.
  Examples: Maximum load test, stress tests, performance under varying temperatures, etc.
- Be sure to include **edge cases** or failure modes (e.g., operating beyond typical conditions) in your scenarios.

Step 3: Create the Test Engineering Matrix
- Develop a **matrix** that outlines the test parameters, methods, and tools to be used. Use the following format:

Test ID	Test Parameter	Test Method/Tool	Pass/Fail Criteria	Responsible Team Member	Number of Runs
1	Load Capacity	Hydraulic Press	Must withstand 1000 N	[Name]	5
2	Heat Resistance	Thermal Chamber	No deformation at 90°C	[Name]	3

Deliverable:
- A **completed Test Engineering Matrix** with at least 5-7 test cases covering critical design parameters.

Part 2: Designing a Test Instance Recording Sheet

Step 1: Format the Test Instance Recording Sheet
- Create a recording sheet for documenting the **results of each individual test run**. Include space for the following fields:

Test ID	Date	Time	Test Run #	Parameter Tested	Result	Observations	Tester
1	10/19/24	10:00 AM	1	Load Capacity	980 N	0.01 in deflection	[Name]

Step 2: Review and Refine
- Ensure the recording sheet provides enough detail to identify any trends or inconsistencies across multiple test runs.
- Share the draft with your team and professor/supervisor for feedback before finalizing.

Deliverable:
- A **Test Instance Recording Sheet** ready to be used during testing.

Part 3: Conducting the Tests

Step 1: Schedule the Tests
- Plan your testing sessions, allowing sufficient time to run each test **a minimum of three times** (ideally 5-7 times) to capture variability.
- Coordinate with team members and ensure the required tools and facilities are booked.

Step 2: Perform the Tests and Record Results
- Run each test according to the procedures outlined in your matrix.
- Use the **Test Instance Recording Sheet** to document the results and any relevant observations. Be meticulous in recording times, conditions, and outcomes.

Step 3: Identify Outliers and Issues During Testing
- During testing, flag any unusual or unexpected results for further investigation.
- If any issues arise with equipment or procedures, document these challenges and adjust your process accordingly.

Part 4: Statistical Analysis and Reporting

Step 1: Calculate Descriptive Statistics
- Use tools like **Excel** or **Python** to calculate key statistical values for each parameter, such as:
  - Mean
  - Standard Deviation
  - Range (Min-Max)
  - Variance

Step 2: Evaluate Consistency
- Assess whether the results are consistent across multiple runs. If large variations exist, determine the likely causes (e.g., equipment variability, operator error, or environmental conditions).

Step 3: Prepare a Statistical Report
- Create graphs (such as bar charts or scatter plots) to visualize the test data.
- Compare results to the pass/fail criteria in the Test Engineering Matrix.

Deliverable:
- A **Statistical Report** summarizing the key results, trends, and any outliers observed during testing.

Part 5: Final Report and Submission

Step 1: Compile Your Findings
- Summarize the entire testing process, including:
  - Overview of test parameters
  - Key observations from the tests
  - Statistical analysis results
  - Identified issues and recommendations for improvements

Step 2: Submit the Final Report
- Organize the following into a **Final Test Engineering Report**:
  - **Test Engineering Matrix**
  - **Test Instance Recording Sheets** (completed)
  - **Statistical Analysis Report**
  - **Recommendations for Design Improvements**

Deliverable:
- A **PDF report** containing all the required components submitted by the assigned deadline.

Timeline and Schedule:
- **Week 1:** Identify test parameters, create the matrix, and finalize the recording sheet
- **Week 2:** Conduct tests and document results
- **Week 3:** Perform statistical analysis and identify trends/issues
- **Week 4:** Compile the final report and submit

Evaluation Criteria:
- **Completeness:** Were all deliverables submitted?
- **Quality of Matrix and Recording Sheets:** Were the test plans well-structured and thorough?
- **Statistical Analysis:** Was the data correctly analyzed and presented clearly?
- **Testing Process:** Were tests conducted methodically and documented properly?
- **Recommendations:** Were any design improvements suggested based on the results?

17 Redesign

17.1 Introduction to Redesign

Redesign is an essential phase in the lifecycle of all engineering design projects. After constructing a working model and conducting rigorous testing, a list of issues, fixes, and areas for improvement is typically generated. The knowledge gained from this testing phase guides the design team in rectifying errors, addressing problems, and implementing enhancements. This iterative process is fundamental to engineering design, ensuring that the final product meets or exceeds the necessary specifications and user requirements.

17.1.1 Iterative Nature of Design

Engineering design is inherently iterative. Redesign represents a critical phase where the initial design is revisited, evaluated, and refined. The goal is to address shortcomings, optimize performance, and ensure the product or process aligns with both the original specifications and any new insights gained during testing. The redesign phase involves several key activities, including:

- **Reviewing Design Specifications**: After testing, the design team should revisit the original specifications to determine if any changes are warranted based on new insights.
- **Analyzing Test Results**: Detailed analysis of test results helps identify specific areas that need attention.
- **Incorporating Feedback**: Interaction with sponsors and customers often provides valuable feedback that should be incorporated into the redesign.
- **User Training Considerations**: Sometimes, improving user training can be an effective way to address certain issues without altering the design.
- **Budget and Resource Review**: Ensuring adequate budget and resources for the redesign activities is crucial.
- **Resolving Unresolved Issues**: Identifying and developing solutions for any outstanding problems.

- **Coordination with Stakeholders**: Working closely with sponsors and other stakeholders to define the scope and objectives of the redesign.
- **Updating the Project Plan**: Revising the project plan to reflect the new scope and activities associated with the redesign.

17.1.2 Continuous Improvement and Practical Constraints

In the real world, redesign is an ongoing process. Designers constantly identify new ways to improve and optimize their products or processes. However, practical constraints such as deadlines and budgets necessitate prioritizing critical issues and optimizing solutions within the available resources. The design team must balance the need for continuous improvement with the realities of project timelines and budget limits.

17.1.3 Capturing Requirements and Enhancements

One of the crucial aspects of the redesign phase is capturing and documenting requirements and feedback from sponsors and customers. This involves:

- **Identifying Valuable Design Features**: Recognizing which features of the design are particularly beneficial to the user and considering enhancing these aspects.
- **Testing and User Feedback**: Conducting additional tests to validate improvements and gathering user feedback to ensure enhanced satisfaction and usability.
- **Addressing Failures and Problems**: Investigating failures and problems to determine their root causes and making necessary changes to prevent recurrence.

17.1.4 Revisiting Design Specifications

The problem definition and design specifications serve as benchmarks for testing and assessment. Redesign activities aim to align the product with these benchmarks. However, as the design evolves, it may become necessary to adapt, change, or add specifications. This adaptive process ensures that the design remains relevant and effective in meeting user needs and expectations.

17.1.5 Iterative Testing and Assessment

Testing, assessment, and redesign form a continuous loop, with each iteration bringing the design closer to the optimal solution. The number of iterations and the degree of improvement achieved depend on the complexity and scope of the design problem. In capstone design projects, where time and resources are limited, the redesign phase must be carefully planned and executed within a constrained timeframe.

17.1.6 Fundamental Design Flaws

In some cases, fundamental or inherent flaws in the design may necessitate a more significant overhaul. This could involve reworking the problem definition and starting the design process anew. Identifying such flaws early can save time and resources in the long run, ensuring that the final product is robust and effective.

17.1.7 Quick Testing and Iterative Refinement

After each redesign iteration, the modified aspects of the design should be promptly tested to verify improvements. This quick feedback loop helps in identifying whether the changes have resolved the issues or introduced new ones. The iterative process continues until the design meets all requirements and performs reliably in real-world conditions.

Redesign is a critical phase in the engineering design process, allowing teams to refine their solutions based on testing and feedback. It involves a structured approach to revisiting and improving the initial design, ensuring that the final product is robust, reliable, and meets all user requirements. By embracing the iterative nature of design and continuously seeking improvements, engineering teams can deliver high-quality solutions that stand the test of time.

17.2 Application of Test Results

The application of test results is a pivotal phase in the engineering design process, particularly in capstone design projects. This phase involves analyzing the outcomes of carefully planned and executed tests to validate the design solution against the established specifications. The results of these tests guide the design team in making informed decisions about necessary modifications, optimizations, and enhancements to the design. By methodically applying test results, the design team can ensure that the final product meets or exceeds all required standards and performs reliably in real-world conditions.

17.2.1 The Role of the Test Matrix

A test matrix is an essential tool that outlines the various tests to be conducted, the parameters to be measured, and the success criteria for each test. For capstone design projects, the test matrix serves as a roadmap for validating the design against its specifications. Each test corresponds to a specific design requirement, such as weight, strength, durability, or functionality.

Table 17-1. Example of a Test Matrix for a Capstone Project.

Test ID	Test Description	Specification	Measurement Method	Success Criteria
1	Weight Measurement	Product weight < 50 lbs	Weighing scale	Weight ≤ 50 lbs
2	Load Bearing Test	Supports weight of 200 lbs	Static load test	No structural failure
3	Durability Test	Operates for 500 hours	Continuous operation	No operational failure
4	Thermal Performance	Max temperature < 70°C	Thermal sensor	Temperature ≤ 70°C
5	Vibration Resistance	Withstands vibrations up to 2g	Vibration table	No structural damage

Each entry in the test matrix includes a unique test ID, a description of the test, the specific design specification being validated, the measurement method, and the success criteria. This structured approach ensures that all critical aspects of the design are systematically evaluated.

17.2.1.1   Analyzing Test Results

17.2.1.1.1   Weight Measurement Example:

If the weight of the product is specified to be less than 50 lbs, the design team weighs the working model using a precise weighing scale. The measured weight is then compared against the specified limit.

- **If the weight exceeds 50 lbs**: The design team must analyze the components contributing to the excess weight. Possible actions include redesigning certain parts to use lighter materials, optimizing the geometry to reduce unnecessary bulk, or reevaluating the assembly process to eliminate redundant components.
- **If the weight is within the limit**: The design team can proceed to the next specification, confident that the weight requirement has been met.

17.2.1.1.2   Iterative Testing and Redesign:

Testing is inherently iterative. If a test reveals that a specification has not been met, the design team revisits the design to identify changes that can bring the product into compliance. This may involve:

- **Changing Materials**: Substituting heavier materials with lighter

17.2 Application of Test Results

alternatives while maintaining or improving strength and durability.
- **Optimizing Geometry**: Redesigning parts to reduce weight without compromising structural integrity or functionality.
- **Reassessing Assembly Methods**: Streamlining the assembly process to eliminate unnecessary parts or steps.

Once modifications are made, the design is retested to verify that the changes have addressed the issue. This cycle continues until all specifications are satisfactorily met.

17.2.2 Addressing Multiple Specifications

Often, multiple specifications must be evaluated and optimized simultaneously. For example, while reducing weight, the design team must ensure that the structural integrity and durability are not compromised. This requires a balanced approach to redesign, where changes are carefully assessed for their impact on all relevant specifications.

Example:
- **Load Bearing and Weight**: While reducing the weight of a structural component, the team must ensure that the component can still support the required load without failure. This might involve using advanced materials like carbon fiber composites that offer high strength-to-weight ratios.
- **Thermal Performance and Durability**: Ensuring that materials used to reduce weight also meet thermal performance specifications, preventing overheating during operation.

17.2.3 Incorporating Sponsor and Customer Feedback

Test results should also incorporate feedback from sponsors and customers. Their practical insights and operational experiences can highlight additional areas for improvement that may not be immediately apparent through standard testing.

Example:
- **User Feedback on Durability**: If users report frequent wear and tear on a specific part, the design team may need to investigate the root cause and consider using more durable materials or redesigning the part for better longevity.
- **Sponsor Requests for Additional Features**: Based on test results, sponsors might request additional features or enhancements. The design team should evaluate the feasibility of these requests and integrate them into the redesign if practical.

17.2.4 Budget and Resource Considerations

The redesign process must also take into account the available budget and resources. Ensuring that the project stays within financial and resource constraints is crucial for successful completion.

Example:
- **Cost-Effective Material Changes**: If a lighter material is required to meet weight specifications but is too expensive, the team must find a cost-effective alternative that balances performance and budget.
- **Resource Allocation**: Ensuring that sufficient time and resources are allocated for retesting after redesigns to verify that all changes meet the required specifications.

17.2.5 Documenting and Updating the Project Plan

As changes are made based on test results, it is essential to document these modifications and update the project plan accordingly. This ensures transparency and provides a clear record of the design evolution.

Steps for Documentation:
1. **Record Test Results**: Document the outcomes of each test, noting whether the specification was met or if modifications are needed.
2. **Detail Redesign Changes**: Describe the specific changes made to the design, including materials used, geometric alterations, and any process modifications.
3. **Update the Project Plan**: Revise the project timeline and milestones to include redesign activities and additional testing phases.

17.2.6 Real-World Implications and Continuous Improvement

In real-world design projects, redesign and testing continue beyond the initial product release. Continuous improvement is a core principle, driven by ongoing feedback and performance data.

Example:
- **Post-Market Testing**: After the product is launched, real-world usage provides valuable data for further refinements. This ongoing process ensures the product remains competitive and meets evolving user needs.
- **Sustainability Considerations**: Modern design increasingly focuses on sustainability. Test results related to environmental impact, such as recyclability and energy efficiency, drive redesigns aimed at reducing the product's ecological footprint.

The application of test results is a critical component of the engineering design process, particularly in capstone projects where student teams must rigorously validate their designs. By systematically analyzing test outcomes, incorporating feedback, and making informed redesign decisions, design teams can ensure that their final product meets all specifications and performs reliably in real-world conditions. This iterative process not only enhances the quality and functionality of the design but also provides invaluable learning experiences for future engineering challenges.

17.3 Adjustments to Design-Build

The testing phase of a design-build project often reveals discrepancies between the actual performance of the working model and the initial design specifications. These discrepancies necessitate adjustments to the design to ensure compliance with the specifications and to improve the overall functionality and performance of the product or process. This section explores the various aspects of making adjustments to the design based on test results, including the iterative nature of the process, dealing with over-constrained systems, and the role of stakeholders in implementing changes.

17.3.1 Iterative Nature of Design Adjustments

Design adjustments are an integral part of the engineering process. They involve revisiting the design based on the feedback from testing and making necessary modifications to improve the design.

17.3.1.1 Steps in the Iterative Adjustment Process:

1. **Identify Discrepancies**: Analyze the test results to identify where the working model deviates from the design specifications.
2. **Determine Root Causes**: Investigate the underlying reasons for these deviations, which could be due to material properties, geometric constraints, assembly methods, or other factors.
3. **Develop Adjustment Strategies**: Formulate strategies to address these discrepancies. This could involve changing materials, altering the geometry, modifying assembly techniques, or incorporating additional features.
4. **Implement Changes**: Make the necessary changes to the design and rebuild the working model.
5. **Retest**: Conduct additional tests to verify that the adjustments have resolved the issues and brought the design into compliance with the specifications.
6. **Document Adjustments**: Keep detailed records of all changes made and the reasons for these changes. This documentation is crucial for future reference and continuous improvement.

Example:
- **Weight Adjustment**: If the working model exceeds the specified weight limit, the team may need to switch to lighter materials or redesign parts to reduce weight while maintaining strength and functionality. After implementing these changes, the team would retest the model to ensure the new weight meets the specification.

17.3.2 Dealing with Over-Constrained Systems

In some cases, it may not be possible to meet all design specifications simultaneously because the system is over-constrained. Over-constrained systems occur when there are conflicting requirements that cannot all be satisfied at the same time.

17.3.2.1 Steps to Address Over-Constrained Systems:

1. **Identify Conflicting Specifications**: Determine which specifications are in conflict and why they cannot be met simultaneously.
2. **Prioritize Specifications**: Work with stakeholders to prioritize the most critical specifications. This prioritization helps in making informed decisions about which constraints can be relaxed.
3. **Modify Specifications**: Adjust or modify the less critical specifications to achieve a viable design. This might involve relaxing certain tolerances or redefining performance criteria.
4. **Redesign and Test**: Implement the changes, rebuild the working model, and conduct tests to verify that the modified design meets the new specifications.

Example:
- **Strength vs. Weight**: If the design requires a part to be both extremely strong and lightweight, but available materials cannot meet both criteria, the team might prioritize strength and accept a slight increase in weight. Alternatively, they could redesign the part to distribute loads more effectively, thereby reducing the need for high-strength materials in some areas.

17.3.3 Role of Stakeholders in Design Adjustments

Engaging all stakeholders—sponsors, customers, team members, and advisors—is crucial when making significant adjustments to the design, especially when design specifications need to be changed.

17.3.3.1 Steps for Stakeholder Engagement:

1. **Communicate Findings**: Clearly communicate the test results and the identified issues to all stakeholders.

2. **Discuss Adjustment Options**: Present the possible adjustment strategies and their implications, including any changes to the design specifications.
3. **Seek Approval**: Obtain stakeholder approval for the proposed changes, ensuring that everyone understands the rationale and agrees on the path forward.
4. **Implement and Review**: Make the approved changes and involve stakeholders in reviewing the results of subsequent tests to ensure that the adjustments have achieved the desired outcomes.

Example:
- **Customer Feedback**: If a customer provides feedback that the product is difficult to use, this feedback should be incorporated into the redesign process. The team might need to adjust ergonomic aspects or add user-friendly features. Engaging the customer in this process ensures that the changes align with their expectations and improve the overall user experience.

17.3.4 Practical Considerations for Design Adjustments

17.3.4.1 Resource Management

- **Budget and Time Constraints**: Ensure that adjustments can be made within the project's budget and timeline. Significant redesigns might require additional resources, so it's essential to plan accordingly.
- **Material Availability**: Check the availability of new materials or components that might be required for adjustments to avoid delays.

17.3.4.2 Risk Management

- **Risk Assessment**: Assess the risks associated with making changes to the design. Consider the potential impact on other parts of the system and ensure that changes do not introduce new issues.
- **Testing Scope**: Ensure that the scope of retesting is comprehensive enough to catch any new problems introduced by the changes.

17.3.4.3 Documentation and Communication

- **Documentation**: Keep detailed records of all adjustments, including the reasons for changes and the results of retests. This documentation is essential for accountability and future reference.
- **Regular Updates**: Provide regular updates to all stakeholders on the progress of adjustments and any new findings from retesting.

Example:
- **Supplier Change**: If the original supplier for a critical component is

unable to meet the required specifications or delivery schedule, the team may need to find an alternative supplier. This change must be documented, and the new components should be thoroughly tested to ensure they meet the necessary standards.

Adjustments to the design-build process are a natural and necessary part of engineering projects. By systematically addressing discrepancies revealed during testing, considering the constraints and priorities of the design, and engaging stakeholders in the adjustment process, design teams can enhance their solutions and ensure that the final product meets or exceeds all required specifications. This iterative approach to design refinement not only improves the quality and performance of the product but also provides invaluable learning experiences for engineering students as they prepare for real-world engineering challenges.

17.4 Major Design Changes

Major design changes become necessary when incremental adjustments and minor specification changes fail to produce a satisfactory design solution. These significant modifications can be challenging and costly, potentially setting back the project timeline considerably. This section explores the reasons for major design changes, the process of implementing these changes, and the importance of stakeholder communication and approval.

17.4.1 Reasons for Major Design Changes

Major design changes are typically prompted by several factors, including:

- **Fundamental Flaws**: The initial design might contain fundamental flaws that cannot be rectified through minor adjustments.
- **Inadequate Performance**: The working model consistently fails to meet critical specifications or performance criteria.
- **New Requirements**: Changes in sponsor or customer requirements that necessitate a complete redesign.
- **Technological Advancements**: New technologies or materials become available that could significantly improve the design.
- **Cost Overruns**: The current design approach might be too costly, necessitating a major change to reduce expenses.

17.4.2 Process of Implementing Major Design Changes

17.4.2.1 Identification and Analysis

- **Root Cause Analysis**: Conduct a thorough analysis to identify the root causes of the design failures.
- **Impact Assessment**: Evaluate the impact of these failures on the overall project goals and objectives.

17.4.2.2 Proposal Development

- **Redesign Options**: Develop several redesign options that could address the identified issues.
- **Feasibility Studies**: Conduct feasibility studies for each redesign option, considering technical, financial, and time constraints.
- **Risk Assessment**: Assess the risks associated with each redesign option, including potential setbacks and new challenges.

17.4.2.3 Stakeholder Communication

- **Detailed Reporting**: Prepare a comprehensive report outlining the need for major design changes, the proposed solutions, and the expected impact on the project.
- **Stakeholder Meetings**: Organize meetings with all stakeholders, including sponsors, customers, and professors, to present the findings and proposed redesigns.
- **Feedback Incorporation**: Collect feedback from stakeholders and incorporate it into the redesign plan.

17.4.2.4 Approval and Documentation

- **Formal Approvals**: Obtain formal approvals from sponsors and professors before proceeding with major changes.
- **Documentation**: Document all decisions, approvals, and changes to maintain a clear record of the redesign process.

17.4.2.5 Implementation

- **Resource Allocation**: Allocate resources, including time, budget, and personnel, to implement the major design changes.
- **Project Plan Update**: Update the project plan to reflect the new timeline, milestones, and deliverables.
- **Redesign Execution**: Execute the redesign according to the updated project plan.

17.4.2.6 Testing and Validation

- **Prototype Development**: Develop new prototypes based on the redesigned solution.
- **Rigorous Testing**: Conduct rigorous testing to ensure that the new design meets all specifications and performance criteria.
- **Iterative Improvements**: Make iterative improvements based on testing results until the design is satisfactory.

17.4.3 Implications of Major Design Changes

17.4.3.1 Project Timeline

- **Delays**: Major design changes can cause significant delays in the project timeline. The design team must account for additional time required for redesign, testing, and validation.
- **Milestones**: Project milestones may need to be adjusted to reflect the new timeline. This could impact the overall project schedule and deadlines.

17.4.3.2 Project Costs

- **Increased Costs**: Major design changes often result in increased costs due to additional materials, labor, and resources required for redesign and testing.
- **Budget Management**: The design team must carefully manage the budget to accommodate these additional costs, potentially seeking additional funding from sponsors.

17.4.3.3 Team Dynamics

- **Morale**: Major setbacks can impact team morale. Effective communication and a clear plan can help maintain team motivation.
- **Roles and Responsibilities**: Redesigning a project may require reassigning roles and responsibilities within the team to leverage individual strengths and expertise.

17.4.3.4 Stakeholder Relations

- **Trust and Confidence**: Keeping stakeholders informed and involved throughout the redesign process helps maintain trust and confidence in the design team's capabilities.
- **Expectations Management**: Clearly communicating the reasons for major changes and the expected outcomes helps manage stakeholder expectations and reduces potential frustration.

17.4.4 Example Scenario

17.4.4.1 Initial Design Challenge

A capstone project team is developing an autonomous delivery robot. The initial design meets most specifications but consistently fails to achieve reliable navigation in various environments, a critical requirement for the project.

17.4.4.2 Root Cause Analysis

The team conducts a root cause analysis and identifies that the navigation system's sensors are inadequate for the range of environments expected. Minor adjustments, such as changing sensor positions and recalibrating software, have failed to solve the issue.

17.4.4.3 Proposed Major Changes

- **New Navigation System**: Propose a new navigation system incorporating advanced LiDAR technology, which is more reliable in diverse environments.
- **Chassis Redesign**: Redesign the robot chassis to accommodate the new sensors and ensure stability across different terrains.
- **Software Overhaul**: Revise the navigation algorithms to integrate with the new sensor data and improve overall system responsiveness.

17.4.4.4 Stakeholder Communication

The team prepares a detailed report and presentation for the sponsors, professor, and other stakeholders. They outline the navigation issues, propose the new design, and provide a revised project plan with updated timelines and costs.

17.4.4.5 Approval and Implementation

Upon receiving approval, the team updates the project documentation, allocates additional budget for the new sensors, and revises the project timeline. The redesign process involves close collaboration with experts in LiDAR technology and iterative testing to validate the new system.

17.4.4.6 Outcome

The new navigation system significantly improves the robot's performance, meeting all critical specifications. Although the project experienced delays and increased costs, the final product is robust and reliable, satisfying the sponsors and demonstrating the team's ability to overcome significant design challenges.

Major design changes are a critical but challenging aspect of the engineering design process. When incremental adjustments and specification changes are insufficient, stepping back and making significant modifications can be necessary. By following a structured process that includes thorough analysis, stakeholder communication, formal approvals, and careful implementation, design teams can navigate these challenges effectively. Although such changes can be costly and time-consuming, they often lead to superior design solutions that meet or exceed initial project objectives.

17.5 Optimization

Optimization in design projects involves determining the set of design parameters that lead to the best possible performance according to a specific objective function. This objective function might aim to minimize weight, time, or cost, or to maximize performance, efficiency, or profit. Design optimization is a critical step in engineering because it helps ensure that a design meets its goals in the most efficient and effective way possible.

17.5.1 Definition of Optimization

Optimization is the process of finding the best solution from all feasible solutions. In the context of engineering design, it involves adjusting the design parameters to achieve the maximum (or minimum) value of an objective function while satisfying all given constraints. The objective function is a quantitative measure of performance, such as weight, cost, efficiency, or strength. Constraints are limitations or requirements that the design must meet, such as material properties, dimensions, or regulatory standards.

- **Objective Function**: A mathematical representation of the goal of the optimization, such as minimizing cost or maximizing efficiency.
- **Design Parameters**: Variables that can be adjusted in the design, such as material type, dimensions, or process parameters.
- **Constraints**: Conditions that the design must satisfy, such as safety regulations, material limits, or budget constraints.

17.5.2 Types of Optimization Problems

Optimization problems can be classified based on the nature of the objective function, the constraints, and the variables involved.

17.5.2.1 Linear vs. Nonlinear Optimization

- **Linear Optimization**: The objective function and constraints are linear functions of the design variables.
- **Nonlinear Optimization**: The objective function or constraints (or both)

are nonlinear functions of the design variables.

17.5.2.2 Constrained vs. Unconstrained Optimization

- **Constrained Optimization**: There are restrictions or bounds on the design variables or the objective function.
- **Unconstrained Optimization**: There are no restrictions on the design variables.

17.5.2.3 Discrete vs. Continuous Optimization

- **Discrete Optimization**: The design variables can take on only discrete values (e.g., integer values, specific material types).
- **Continuous Optimization**: The design variables can take on any value within a continuous range.

17.5.3 Optimization Methods

Optimization methods can be broadly categorized into traditional mathematical techniques and modern computational methods.

17.5.3.1 Traditional Mathematical Techniques

- **Gradient-Based Methods**: These methods use the gradient of the objective function to find the optimal solution. Examples include the Gradient Descent Method and Newton's Method.
- **Linear Programming (LP)**: A method for optimizing a linear objective function subject to linear equality and inequality constraints.
- **Quadratic Programming (QP)**: An extension of LP where the objective function is quadratic, and the constraints are linear.

17.5.3.2 Computational Methods

- **Genetic Algorithms**: These are search heuristics that mimic the process of natural evolution. They are particularly useful for solving complex optimization problems with large search spaces.
- **Simulated Annealing**: A probabilistic technique that explores the search space and avoids local minima by allowing occasional increases in the objective function value.
- **Particle Swarm Optimization (PSO)**: This method simulates the social behavior of birds flocking or fish schooling to find optimal regions of the search space.
- **Artificial Neural Networks (ANNs)**: ANNs can be trained to approximate the objective function and find optimal solutions.

17.5.4 Optimization in Design Process

17.5.4.1 Problem Formulation

The first step in optimization is to clearly define the problem. This involves specifying the objective function, identifying the design variables, and listing all constraints.

Example: Suppose we need to design a lightweight yet strong beam for a bridge. The objective function could be to minimize the weight of the beam, with constraints on its strength and dimensions.

17.5.4.2 Model Development

Develop a model that describes the relationship between the design variables and the objective function. This model can be analytical, numerical, or empirical.

Example: The model could include equations that relate the material properties and dimensions of the beam to its weight and strength.

17.5.4.3 Solution Search

Use appropriate optimization techniques to search for the best set of design variables. This involves running simulations, performing calculations, or using software tools.

Example: Use a genetic algorithm to explore different combinations of materials and dimensions to find the lightest beam that meets the strength requirements.

17.5.4.4 Validation

Validate the optimal solution through testing or further simulations to ensure that it meets all constraints and performs as expected.

Example: Build a prototype of the optimized beam and test it under load to verify its strength and weight.

17.5.5 Practical Application of Optimization

17.5.5.1 Example: Minimizing Weight in Aerospace Design

In aerospace design, minimizing weight is crucial due to its direct impact on fuel efficiency and performance. An optimization problem might involve finding the lightest possible structure for an aircraft wing that can withstand all operational loads.

- **Objective Function**: Minimize the weight of the wing.
- **Design Variables**: Material type, thickness of the skin, number of ribs, and their placement.
- **Constraints**: Must meet strength and stiffness requirements,

aerodynamic performance, and regulatory standards.
- **Optimization Method**: Use a combination of gradient-based methods for initial exploration and genetic algorithms for refining the design.
- **Outcome**: The optimal design will use advanced composite materials and an optimized rib structure to achieve minimum weight while meeting all performance criteria.

17.5.6 Challenges in Optimization

17.5.6.1 Complexity and Computation

Many optimization problems are highly complex and require significant computational resources. The design team must balance the need for detailed models with the available computational power.

17.5.6.2 Multiple Objectives

In real-world applications, there are often multiple objectives that need to be optimized simultaneously, such as cost, performance, and durability. This requires multi-objective optimization techniques, which can handle trade-offs between conflicting objectives.

17.5.6.3 Uncertainty and Variability

Design parameters and constraints often involve uncertainties, such as variations in material properties or operating conditions. Robust optimization techniques can help find solutions that perform well under a range of conditions.

17.5.6.4 Dynamic and Real-Time Optimization

Some applications, such as adaptive systems and real-time controls, require dynamic optimization where the optimal solution changes over time. This requires fast, adaptive algorithms capable of real-time computation.

17.5.7 Optimization Software Tools

Several software tools are available to assist with optimization in engineering design. These tools provide powerful capabilities for modeling, simulation, and optimization, making them essential for complex design projects.

17.5.7.1 Popular Optimization Tools:
- **MATLAB**: Widely used for numerical computation, simulation, and optimization.
- **ANSYS**: Provides advanced simulation and optimization capabilities for

engineering designs.
- **SolidWorks**: Includes optimization tools for mechanical design.
- **Autodesk Inventor**: Offers integrated optimization features for product development.
- **COMSOL Multiphysics**: Allows for the optimization of multiphysics models.

Optimization is a fundamental aspect of engineering design, enabling designers to achieve the best possible performance for their products and processes. By systematically adjusting design parameters and using advanced optimization techniques, engineers can find solutions that meet or exceed all specifications while considering constraints and real-world challenges. Through careful problem formulation, model development, and solution search, optimization leads to innovative, efficient, and effective designs. As the complexity of engineering problems continues to grow, the role of optimization in ensuring high-quality, reliable, and cost-effective solutions becomes increasingly important.

17.6 Quality Engineering

The concept of quality is multifaceted and often varies across different fields and product classes. In general, quality can be understood as the degree to which a product or service is free of defects, weaknesses, and deficiencies, and meets or exceeds customer expectations. This section explores the various dimensions of quality as proposed by Garvin, namely performance, features, reliability, conformance, durability, serviceability, aesthetics, and perceived quality, and discusses how these dimensions apply to engineering design.

17.6.1　Garvin's Eight Dimensions of Quality

17.6.1.1　Performance

Performance describes a product's operating characteristics and is an objective and measurable aspect of quality. For example, in the context of a car, performance includes factors such as acceleration, braking, handling, top speed, and comfort. In engineering design, performance metrics are often specified as part of the design requirements and are rigorously tested to ensure that the product meets these specifications.

Example: For a high-performance vehicle, performance metrics might include achieving a certain acceleration from 0 to 60 mph, maintaining stability at high speeds, and providing a comfortable ride over various road conditions.

17.6 Quality Engineering

17.6.1.2 Features

Features are additional aspects of a product that provide extra benefits or capabilities. These can include both aspects explicitly requested by the sponsor and those added by the design team to enhance the product's value. Features distinguish one design solution from another and provide sponsors with desirable options.

Example: In a smartphone, features might include a high-resolution camera, facial recognition, extended battery life, and water resistance. These features add value beyond the basic functionality of making calls and sending messages.

17.6.1.3 Reliability

Reliability is the probability that a product or process will function without failure over a specified period. It can be quantified using metrics such as mean time between failures (MTBF) or mean time to first failure (MTTF). Reliability testing involves subjecting the product to conditions that simulate its intended use and measuring how often failures occur.

Example: For a medical device, reliability might be measured by the number of successful operations before any failure occurs, ensuring that the device performs consistently under various conditions.

17.6.1.4 Conformance

Conformance measures how well a product or process adheres to established standards, specifications, or regulations. It can be assessed through yield rates during manufacturing, where a high yield rate indicates high conformance and low defect rates. Conformance ensures that the product meets industry standards and customer expectations.

Example: In the manufacturing of electronic components, conformance might involve adhering to industry standards for voltage, current, and resistance to ensure compatibility and safety.

17.6.1.5 Durability

Durability refers to the product's useful service life before it deteriorates to the point where it requires replacement or significant repair. A durable product continues to perform as expected over a long period, reducing the need for frequent repairs and lowering the total cost of ownership.

Example: For construction equipment, durability might be measured by the number of operating hours before major maintenance is required, indicating the equipment's robustness in harsh working conditions.

17.6.1.6 Serviceability

Serviceability encompasses the ease and speed with which a product can be repaired and returned to operational status. It includes the availability of spare parts, the simplicity of disassembly and reassembly, and the effectiveness of customer service. High serviceability ensures minimal downtime and high customer satisfaction.

Example: For consumer electronics, serviceability might involve providing easily accessible repair manuals, offering extended warranties, and maintaining a network of authorized repair centers.

17.6.1.7 Aesthetics

Aesthetics pertains to the subjective sensory attributes of a product, such as its appearance, feel, sound, and smell. While aesthetics are subjective, they play a crucial role in customer satisfaction and can significantly influence purchasing decisions.

Example: In automotive design, aesthetics might include the car's exterior design, the quality of interior materials, and the sound of the engine, all contributing to the overall appeal of the vehicle.

17.6.1.8 Perceived Quality

Perceived quality is the customer's perception of a product's overall quality, often influenced by brand reputation, marketing, and user reviews. It is an indirect measure of a product's value and usability and can significantly impact customer satisfaction and loyalty.

Example: In the consumer electronics market, perceived quality can be influenced by online reviews, brand reputation, and product endorsements, shaping customer expectations and experiences.

17.6.2 Application of Quality Engineering in Design

Quality engineering integrates these dimensions into the design process to ensure that the final product meets or exceeds customer expectations. This involves several key practices:

17.6.2.1 Quality Planning

Quality planning involves identifying customer needs and defining the quality objectives and specifications that the product must meet. This includes setting measurable performance targets and establishing criteria for features, reliability, conformance, durability, serviceability, aesthetics, and perceived quality.

Example: For a new wearable fitness tracker, quality planning might involve defining battery life, waterproof rating, accuracy of health metrics, and user interface design based on customer feedback and market research.

17.6.2.2 Quality Assurance

Quality assurance (QA) encompasses the systematic activities implemented in a quality system to ensure that quality requirements for a product or service are fulfilled. QA involves process planning, validation, verification, and regular inspections to ensure that all stages of the design and manufacturing process meet the specified standards.

Example: In software development, QA might include code reviews, automated testing, and user acceptance testing to ensure the software functions correctly and meets user expectations.

17.6.2.3 Quality Control

Quality control (QC) involves the operational techniques and activities used to fulfill quality requirements. This includes the inspection and testing of materials, components, and finished products to identify defects and ensure conformance to specifications.

Example: In the production of automotive parts, QC might involve using precision measurement tools to inspect the dimensions and tolerances of machined components, ensuring they meet design specifications.

17.6.2.4 Continuous Improvement

Continuous improvement is the ongoing effort to enhance products, services, and processes to increase efficiency, reduce waste, and improve quality. Techniques such as Six Sigma, Lean Manufacturing, and Total Quality Management (TQM) are often employed to drive continuous improvement.

Example: In a manufacturing plant, continuous improvement might involve implementing Lean principles to streamline production workflows, reduce lead times, and minimize defects.

17.6.3 Tools and Techniques for Quality Engineering

17.6.3.1 Statistical Process Control (SPC)

SPC involves using statistical methods to monitor and control a process, ensuring that it operates at its full potential. Control charts, process capability analysis, and other statistical tools are used to identify variations and ensure process stability.

Example: In pharmaceutical manufacturing, SPC might be used to monitor the consistency of tablet weights and drug concentrations, ensuring they stay within specified limits.

17.6.3.2 Failure Mode and Effects Analysis (FMEA)

FMEA is a systematic approach for identifying potential failure modes, assessing their impact, and prioritizing actions to mitigate risks. It helps in preventing defects and improving product reliability.

Example: In aerospace engineering, FMEA might be used to analyze the potential failure modes of critical components, such as engines or control systems, and implement design changes to enhance reliability.

17.6.3.3   Design of Experiments (DOE)

DOE is a statistical method used to plan, conduct, and analyze controlled tests to evaluate the factors that affect the performance of a product or process. It helps in identifying the optimal settings for design parameters to achieve the desired quality.

Example: In chemical process engineering, DOE might be used to determine the optimal temperature, pressure, and catalyst concentration to maximize yield and purity in a chemical reaction.

17.6.3.4   Root Cause Analysis (RCA)

RCA is a problem-solving method used to identify the underlying causes of defects or failures. Techniques such as the 5 Whys, Fishbone Diagram, and Fault Tree Analysis are commonly used in RCA.

Example: In automotive manufacturing, RCA might be used to investigate the root cause of a recurring defect in the assembly line, leading to process improvements to eliminate the issue.

Quality engineering is essential in ensuring that products and processes meet the highest standards of performance, reliability, and customer satisfaction. By focusing on Garvin's eight dimensions of quality and employing robust quality engineering practices, design teams can create products that not only meet specifications but also exceed customer expectations. Through quality planning, assurance, control, and continuous improvement, engineers can deliver superior products and services that stand out in the competitive marketplace.

17.7 Cost Evaluation

Cost evaluation is a critical component of the engineering design process. It involves assessing the total expenses associated with a project to ensure that it remains within budget and delivers value to the stakeholders. When a redesign is necessary, the cost evaluation becomes even more crucial, as it can significantly impact the financial feasibility of the project. This section delves into the importance of cost evaluation, the elements that must be considered, and the tools and methods used to perform a thorough cost assessment.

17.7 Cost Evaluation

17.7.1 Importance of Cost Evaluation

1. **Budget Adherence**: Ensuring that the project stays within the allocated budget is fundamental to its success. Cost overruns can jeopardize the project's completion and lead to financial losses.
2. **Resource Allocation**: Proper cost evaluation helps in the efficient allocation of resources, ensuring that funds are used effectively to achieve the project goals.
3. **Decision Making**: Cost evaluation provides essential data for making informed decisions regarding redesigns, resource adjustments, and overall project management.
4. **Stakeholder Confidence**: Accurate cost assessments and adherence to budgets enhance stakeholder confidence and support for the project.

17.7.2 Elements of Cost Evaluation

1. **People**
   - **Students**: Time spent by students on the project should be quantified. This includes the hours dedicated to research, design, prototyping, testing, and documentation.
   - **Consultants**: Fees for external consultants who provide specialized expertise.
   - **Faculty Advisors**: Time contributed by faculty advisors, which might be calculated based on their hourly rate or salary.
   - **Mentors and Sponsors**: Time and resources provided by company staff and mentors associated with the project.

2. **Facilities**
   - **Laboratories**: Usage of university or sponsor laboratories, including electricity, water, and other utilities.
   - **Workshops and Maker Spaces**: Costs associated with using workshops, including maintenance and consumables.

3. **Equipment**
   - **3D Printers**: Depreciation, maintenance, and materials for 3D printing.
   - **Testing Equipment**: Usage and maintenance costs for testing equipment and instruments.
   - **Software Licenses**: Costs for software tools used in design, simulation, and analysis (e.g., CAD software, simulation tools).

4. **Materials**
   - **Raw Materials**: Costs of metals, plastics, composites, and other raw materials used in prototyping and building the design.
   - **Purchased Components**: Costs of off-the-shelf components,

fasteners, adhesives, and other consumables.
- **Shipping and Handling**: Expenses for shipping materials and components to the project site.

5. **Miscellaneous Costs**
   - **Travel**: Expenses for travel to sponsor sites, conferences, or other relevant locations.
   - **Documentation**: Costs of printing reports, posters, and other documentation materials.
   - **Overhead**: Indirect costs associated with project management, administrative support, and other non-direct expenses.

17.7.3 Tools and Methods for Cost Evaluation

17.7.3.1 Project Management Software

Tools like MS Project or Primavera can help in tracking and evaluating costs throughout the project lifecycle. These tools allow for detailed cost accounting and reporting, making it easier to identify cost overruns and areas for potential savings.

17.7.3.2 Cost Estimation Models

Several cost estimation models can be used, including:
- **Analogous Estimating**: Using historical data from similar projects to estimate costs.
- **Parametric Estimating**: Using statistical relationships between historical data and other variables to estimate costs.
- **Bottom-Up Estimating**: Breaking down the project into smaller components and estimating the cost for each component, then summing these estimates to get a total project cost.

17.7.3.3 Earned Value Management (EVM)

EVM is a project management technique that integrates scope, time, and cost data to assess project performance and progress. It helps in identifying variances between planned and actual costs and schedules.

17.7.3.4 Cost-Benefit Analysis (CBA)

CBA involves comparing the costs of a project or redesign with the expected benefits. This analysis helps in determining whether the project is financially viable and if the redesign will provide a return on investment.

17.7.3.5 Life-Cycle Cost Analysis (LCCA)

LCCA considers all costs associated with a product over its entire life cycle, from initial design and manufacturing through to maintenance and disposal. This method provides a comprehensive view of the total cost of ownership.

17.7.4 Impact of Redesign on Cost Evaluation

Redesign activities can significantly influence the overall cost of a project. When planning for a redesign, it is essential to account for the following:

17.7.4.1 Additional Resources

- **Materials and Components**: Additional or replacement materials and components needed for the redesign.
- **Labor**: Extra hours required by the design team, consultants, and advisors to complete the redesign.
- **Equipment and Facilities**: Additional usage of equipment and facilities for testing and prototyping.

17.7.4.2 Time

- **Extended Project Duration**: Redesigns can extend the project timeline, leading to increased costs associated with labor, facilities, and equipment usage.
- **Opportunity Costs**: Delays in project completion may lead to missed opportunities or additional costs associated with late delivery.

17.7.4.3 Risk and Contingency Planning

- **Risk Mitigation**: Costs associated with mitigating risks identified during the redesign phase.
- **Contingency Funds**: Allocating funds for unforeseen expenses that may arise during the redesign process.

17.7.5 Example of Cost Evaluation

Consider a capstone project where the design team is developing a new type of energy-efficient light bulb. The initial budget is $10,000, but after testing the prototype, several issues are identified, necessitating a redesign.

Initial Costs:
- **Materials**: $2,500
- **Labor (Students)**: $3,000
- **Consultants**: $1,000
- **Equipment Usage**: $1,500

- **Facilities**: $1,000
- **Miscellaneous**: $1,000
  Total Initial Costs: $10,000

Redesign Costs:
- **Additional Materials**: $1,500
- **Additional Labor (Students)**: $2,000
- **Consultants (Additional Hours)**: $500
- **Extended Equipment Usage**: $1,000
- **Additional Facilities Usage**: $500
- **Miscellaneous**: $500

Total Redesign Costs: $6,000
New Total Costs: $16,000

In this example, the total cost of the project increased by 60% due to the redesign. The design team must justify these additional expenses to the sponsors and ensure that the benefits of the redesign (e.g., improved energy efficiency, longer product life) outweigh the increased costs.

Cost evaluation is an integral part of the engineering design process, ensuring that projects are completed within budget and provide value to stakeholders. By systematically accounting for all resources used and incorporating cost assessments into the project plan, design teams can manage expenses effectively and make informed decisions about redesigns. Tools like project management software, cost estimation models, EVM, CBA, and LCCA are essential for thorough cost evaluation. In capstone design projects, accurate cost evaluation reflects the maturity and completeness of the design work, ultimately contributing to the project's success.

17.8 Return on Investment

Implementing new design solutions or process improvements in manufacturing often requires significant capital expenditure. Evaluating the Return on Investment (ROI) is crucial for determining the economic feasibility of such changes. ROI analysis helps in understanding how long it will take for the savings generated by the new process to offset the initial investment and, consequently, when the investment will start generating profits.

17.8.1 Economic Analysis and Future Value of Money

The ROI concept involves evaluating the future value of money, similar to the analysis used for loans and investments. The future value calculations consider both simple and compound interest.

17.8.1.1 Simple Interest

Simple interest is calculated using the formula:

$$FV = I \times (1 + (R \times T))$$

where:
- FV is the future value.
- I is the initial investment.
- R is the annual interest rate.
- T is the number of years.

For example, if $50,000 is invested at an annual simple interest rate of 5% for five years, the future value is:

$$FV = 50{,}000 \times (1 + (0.05 \times 5)) = 50{,}000 \times 1.25 = 62{,}500$$

17.8.1.2 Compound Interest

Compound interest is calculated using the formula:

$$FV = I \times (1 + R)^T$$

where:
- FV is the future value.
- I is the initial investment.
- R is the annual interest rate.
- T is the number of years.

Using the same example, the future value with compound interest is:
$$FV = 50{,}000 \times (1 + 0.05)^5 = 50{,}000 \times 1.2762816 = 63{,}814.08$$

17.8.1.3 Calculating Payback Period

To determine the payback period for an investment, we need to calculate how long it will take for the savings or additional profits generated by the new process to equal the initial investment.

Suppose the investment of $50,000 results in annual savings or additional profits of $5,000. The payback period can be calculated using the following formula for the payment P :

$$P = \frac{R \times PV}{1 - (1 + R)^T}$$

where:
- P is the annual payment (savings or profits).
- PV is the present value (initial investment).
- R is the annual interest rate.
- T is the number of years.

Plugging in the values:

$$5{,}000 = \frac{0.05 \times 50{,}000}{1 - (1 + 0.05)^T}$$

Solving for T:

$$5{,}000 = \frac{2{,}500}{1 - (1.05)^T}$$

Rearranging:

$$1 - (1.05)^{-T} = \frac{2{,}500}{5{,}000} = 0.5$$

$$(1.05)^{-T} = 0.5$$

Taking the natural logarithm of both sides:

$$-T \ln(1.05) = \ln(0.5)$$

Solving for T:

$$T = -\frac{\ln(0.5)}{\ln(1.05)} \approx 14.21$$

Thus, the payback period is approximately 14.21 years.

17.8.1.4 General Formula for Payback Period

The general formula to solve for T can be expressed as:

17.8 Return on Investment

$$T = -\frac{\ln\left[1 - \frac{PV \times R}{P}\right]}{\ln[1+R]}$$

where:
- PV is the present value (initial investment).
- R is the annual interest rate.
- P is the annual savings or profit.
- T is the number of years.

Using this formula, we can determine the payback period for any investment based on the given parameters.

17.8.2 Practical Application

17.8.2.1 Example: Investment in a New Manufacturing Process

Let's consider a company investing $100,000 in a new manufacturing process that is expected to save $10,000 annually. The interest rate is 4%.

1. Calculate the payback period:
2.

$PV = 100,000$
$R = 0.04$
$P = 10,000$

Using the formula:

$$T = -\frac{\ln\left[1 - \frac{100,000 \times 0.04}{10,000}\right]}{\ln[1+0.04]}$$

$$T = -\frac{\ln[1-4]}{\ln[1.04]}$$

Since $1 - 4$ results in a negative value, we realize that our scenario is improperly scaled for direct payback computation. Instead, let's consider a scenario where P and PV yield a realistic period calculation.

Corrected Example:

Let's consider $50,000 investment saving $7,000 annually:

$$PV = 50,000$$

$$R = 0.04$$

$$P = 7,000$$

$$T = -\frac{\ln\left[1 - \frac{50,000 \times 0.04}{7,000}\right]}{\ln[1.04]}$$

$$T = -\frac{-0.3365}{0.0392} \approx 8.59$$

Thus, the payback period is approximately 8.59 years.

Evaluating the ROI is crucial for assessing the feasibility and profitability of new design solutions and process improvements. By understanding the time required to recover the initial investment and begin generating profits, companies can make informed decisions about where to allocate their resources. The mathematical foundation provided here ensures that these decisions are based on solid economic principles, helping to maximize the return on capital investments.

17.9 Design Optimization

Design optimization is a critical component of engineering design and problem-solving. In the course of developing a solution to a design problem, engineers often encounter multiple possible solutions, each with varying degrees of effectiveness and efficiency. The primary goal of design optimization is to refine the current design solution to achieve better performance, cost efficiency, and compliance with specifications.

17.9.1 Purpose of Design Optimization

The objective of design optimization is to find the best possible solution among a range of potential solutions. This involves identifying and implementing changes that enhance the design's performance, reduce costs, improve reliability, or achieve other desired outcomes. Optimization is an iterative process that involves continuous evaluation and refinement of the design.

17.9.2 Methods of Design Optimization

Several optimization methods can be employed in the design process. These methods range from mathematical and numerical techniques to more practical, intuitive approaches that are well-suited for capstone design projects.

17.9.2.1 Mathematical and Numerical Methods

Mathematical models for design performance functions can be created to apply optimization methods. These models often use linear or non-linear optimization techniques. Linear optimization is simpler and more commonly used, while non-linear optimization is more complex and typically reserved for special cases. These methods involve defining an objective function that quantifies the performance of the design in terms of specific parameters.

17.9.2.2 Practical Design Optimization Methods

More practical methods, such as testing, improvements, trial-and-error, and intuitive approaches, are often more applicable to capstone design projects. These methods do not always require complex mathematical modeling and can be highly effective in iterative design refinement.

Testing and User Trials

Testing and user trials are fundamental to design optimization. By conducting tests and collecting user feedback, design teams can identify weaknesses, deficiencies, failures, and errors in the design. This information can then be used to make targeted improvements, optimizing the design based on empirical evidence.

Trial-and-Error

Trial-and-error is a practical and intuitive method of design optimization. This approach involves making incremental changes to the design and evaluating their impact on performance. If a change results in improved performance, it is retained; if not, it is discarded. This iterative process continues until the optimal design is achieved.

Numerical Algorithms and Methods

Numerical algorithms can also be employed for design optimization. These methods involve defining an objective function that represents various design features such as cost, performance, weight, strength, thermal output, and safety factors. The objective function is typically a vector function, with each component representing a specific aspect of the design. The design parameters that can be controlled and modified by the designer are represented as variables in the objective function.

17.9.3 Objective Function and Design Parameters

The objective function represents various aspects of the design, such as cost, performance, weight, strength, thermal output, and safety factors. The design parameters that can be controlled and modified by the designer might include material choices, dimensions, configurations, loads, temperatures, flow rates, accelerations, speeds, and process activity times.

For example, in a product design project, the objective function might include components such as:
- **Cost**: The total cost of materials and manufacturing.
- **Performance**: Measures of how well the product meets functional requirements.
- **Weight**: The overall weight of the product.
- **Strength**: The mechanical strength of the product.
- **Thermal Output**: The amount of heat generated by the product.
- **Factor of Safety**: A measure of the design's safety margin.

17.9.4 Managing the Optimization Process

To manage the optimization process effectively, it is essential to keep the number of design parameters relatively small. This makes the problem more manageable and allows for a more focused optimization effort. Additionally, the characteristics of the objective function with respect to the design parameters may be non-linear, which can complicate the optimization process. Keeping parametric variations small can help treat the optimization problem as quasilinear, simplifying the analysis.

17.9.5 Practical Example: Capstone Design Optimization

Consider a capstone project involving the design of a lightweight, energy-efficient drone. The initial design meets basic functional requirements but is heavier than desired and consumes more power than specified. The design team aims to optimize the drone for weight and energy efficiency.

Step 1: Define Objective Function

The objective function for this optimization might include components such as weight, power consumption, and cost, with each represented as a function of design parameters like material selection, component dimensions, and battery capacity.

Step 2: Identify Design Parameters

The design parameters to be optimized might include:
- Material type for the frame (e.g., aluminum, carbon fiber)
- Dimensions of the frame and propellers

- Battery capacity and type
- Motor specifications

Step 3: Implement Changes and Evaluate

Using a trial-and-error approach, the design team might experiment with different materials for the frame. For instance, switching from aluminum to carbon fiber could significantly reduce weight but increase cost. The team tests this change and finds that the weight reduction is substantial, and the performance improves within acceptable cost limits.

Next, the team explores optimizing the propeller dimensions. Increasing the diameter of the propellers might reduce power consumption by improving aerodynamic efficiency. This change is tested, and the results show a noticeable improvement in energy efficiency.

Step 4: Numerical Optimization

The team uses numerical optimization methods to fine-tune the design parameters. For instance, they might use software to model the drone's performance based on different combinations of material types, dimensions, and motor specifications. The software runs simulations to identify the combination that maximizes energy efficiency while keeping weight and cost within acceptable limits.

Step 5: Review and Iterate

The optimized design is reviewed, and further adjustments are made based on testing results and user feedback. This iterative process continues until the drone meets all specified requirements for weight, power consumption, and cost.

Design optimization is a crucial part of engineering design, enabling teams to refine their solutions to achieve the best possible outcomes. By employing a combination of practical methods, such as trial-and-error and testing, along with numerical optimization techniques, designers can systematically improve their designs. In capstone projects, this process not only leads to better design solutions but also provides valuable learning experiences for engineering students, preparing them for real-world engineering challenges.

17.11 Assignments

Assignment 17-1: Creating a Test Engineering Matrix and Running Statistical Tests

Objective:

This assignment challenges your design team to **apply test results to redesign your prototype or product**, correcting errors, addressing deficiencies, and implementing improvements. The goal is to demonstrate how iterative design improves product performance and aligns with design specifications. Additionally, you will reflect on the design process, document all redesign steps, and make recommendations for future improvements.

Assignment Instructions:

Part 1: Analyze Test Results and Identify Design Deficiencies

  Step 1: Review Test Results
  - Compile the **results from your Test Engineering Matrix and Test Instance Recording Sheets**.
  - Identify **parameters or performance areas** where the design failed to meet the specifications or user requirements.

  Deliverable:
  - A **summary of test results** highlighting any errors, performance gaps, or deviations from expected outcomes.

  Step 2: Root Cause Analysis
  - For each identified issue, conduct a **root cause analysis** to determine why the failure or deficiency occurred.
  - Use methods such as **5 Whys** or **Fishbone Diagrams** to explore possible causes (e.g., material issues, poor tolerances, incorrect assembly, user error).

  Deliverable:
  - A **list of identified root causes** and the impact of each on the design's performance.

Part 2: Plan the Redesign Iteration

Step 1: Prioritize Issues for Redesign
- Not all issues may require immediate redesign. Prioritize issues based on **severity, customer impact, and feasibility** for correction within the given timeframe.
- Assign **responsibility** for each issue to specific team members.

Step 2: Brainstorm Design Improvements
- Generate **multiple solutions** to address each identified issue (e.g., change materials, modify dimensions, adjust tolerances).
- Evaluate each solution's **feasibility, cost, time requirements, and potential impact** on other aspects of the design.

Deliverable:
- A **Redesign Plan** listing all proposed improvements, responsible team members, and timeline for implementation.

Part 3: Implement the First Redesign Iteration

Step 1: Document Design Changes
- For each change, create **detailed design sketches, CAD models, or updated schematics**. Ensure that these changes are recorded with version control (e.g., "Version 2.0").
- Include explanations of why the specific change was chosen and how it addresses the identified issue.

Step 2: Build the Modified Prototype or Product
- Incorporate the design changes into your **prototype** using updated materials, dimensions, or components.

Step 3: Re-Test the Modified Design
- Use the **same testing methods and parameters** as the initial tests to ensure comparability.
- **Run each test multiple times** (3 to 5 times) and document the results in your Test Instance Recording Sheet.

Deliverable:
- Updated **test results** for the redesigned product.

Step 4: Document the Redesign Iteration
- Create a **Redesign Iteration Log** detailing:
  - **What changes were made**
  - **Why the changes were necessary**
  - **Testing outcomes** after redesign
  - **New issues or observations** uncovered during the second round of testing

---

Part 4: Plan and Execute Additional Redesign Iterations (if needed)

Step 1: Assess Results from First Iteration
- Review the updated test results. If the design still does not meet all specifications, **plan further redesigns** as necessary.

Step 2: Iterate and Test
- Repeat the **design-build-test-redesign cycle** until the product meets the required specifications.
- Each iteration should aim to **solve new or persistent issues** while maintaining a focus on improving overall performance.
- Document **every change and result** thoroughly in the Redesign Iteration Log.

Deliverable:
- **Complete Redesign Iteration Log** with a record of all iterations made.

Part 5: Assess the Entire Design Process and Make Recommendations

Step 1: Assess the Effectiveness of the Redesign Process
- Evaluate the **overall effectiveness of your design changes**. Did the redesign achieve the intended performance improvements?
- Identify **any design trade-offs** made during the redesign process (e.g., higher cost for better performance).

Step 2: Reflect on Lessons Learned
- Document what **lessons were learned** during the redesign and testing

process.
- **What worked well?**
- **What challenges did you encounter?**
- **What could have been done differently?**

Step 3: Recommend Future Design Improvements
- o Provide **recommendations for future improvements** based on your experience.
- o Answer: **If you could restart your design from the beginning, what would you do differently?**

Deliverable:
- o A **Redesign Process Assessment Report** (500-800 words) that:
  - Summarizes your design challenges and successes
  - Reflects on the lessons learned from the entire process
  - Provides actionable recommendations for future projects

Part 6: Compile the Final Report

Step 1: Organize Your Report
- o Compile all **design sketches, iteration logs, test results, and process assessments** into a final report.
- o Ensure the report is well-organized and easy to follow, with clear section headings for each part of the assignment.

Step 2: Submit Your Final Report
- o Submit the final report as a **PDF document** along with any supplemental files (e.g., CAD models, spreadsheets with test data).

Deliverable:
- o **Final Redesign Report** containing:
  - Test result summary
  - Redesign iteration logs
  - Final test results
  - Process assessment and recommendations

Timeline and Suggested Schedule:
- **Week 1:** Analyze test results and plan redesign

- **Week 2:** Implement and test the first redesign iteration
- **Week 3:** Assess results, make further changes (if needed), and complete additional iterations
- **Week 4:** Finalize process assessment and submit the final report

---

Evaluation Criteria:
- **Completeness:** Were all parts of the assignment completed and documented?
- **Design Iteration Quality:** Were redesigns meaningful and based on clear test results?
- **Documentation:** Were all design steps, tests, and redesigns properly documented?
- **Process Reflection:** Did the team provide thoughtful reflections and recommendations for future improvements?
- **Professionalism:** Is the final report well-organized and polished?

18   Other Considerations

Engineering design is a discipline that extends far beyond the realm of pure functionality and innovation. As we stand at the cusp of unprecedented technological advancements, the responsibility of engineers to consider the broader implications of their designs has never been more critical. This chapter delves into eight fundamental considerations that should be at the forefront of every engineer's mind throughout the design process: safety, ergonomics, health impacts, environmental impacts, social impacts, ethical considerations, political considerations, and sustainability.

These factors are not mere afterthoughts or regulatory checkboxes; they are integral components that shape the very essence of good design. By thoroughly examining each of these aspects, we aim to cultivate a holistic approach to engineering that not only solves immediate problems but also contributes positively to society and the world at large.

Safety forms the cornerstone of responsible design, transcending industries and applications. It serves as the fundamental principle upon which all other considerations are built. Engineers must be adept at identifying potential hazards, evaluating their likelihood and severity, and implementing appropriate mitigation strategies. This process involves not only considering obvious risks but also anticipating unlikely scenarios that could have catastrophic consequences.

Ergonomics focuses on optimizing the interaction between humans and the systems they use. This field is crucial for creating designs that are not only functional but also comfortable, efficient, and safe for human use. It encompasses both physical and cognitive aspects, ensuring that our creations align with human capabilities and limitations.

The health impacts of engineering designs extend beyond immediate safety concerns to long-term effects on human wellbeing. This includes considerations of both physical and mental health, as well as the potential for designs to actively promote healthier lifestyles.

In an era of climate change and environmental degradation, considering the ecological impact of engineering designs is more critical than ever. Engineers must strive to minimize negative environmental effects and even contribute positively to ecosystem health. This involves analyzing the entire life cycle of a product or system and embracing principles of circular economy design.

The social impacts of engineering designs are profound and far-reaching. Our creations do not exist in isolation; they are intrinsically linked to the social fabric of our communities. Engineers must consider how their decisions can influence social structures, cultural norms, and individual behaviors, striving to create solutions that are culturally appropriate and meet the actual needs of diverse communities.

Ethical considerations form the moral backbone of engineering practice. Engineers must navigate complex ethical dilemmas, balancing various stakeholder interests while holding paramount the safety, health, and welfare of the public. This becomes increasingly challenging in the face of rapidly evolving technologies with far-reaching implications.

Political factors, though often overlooked, can significantly influence the viability and implementation of engineering designs. Engineers must navigate a complex regulatory landscape, anticipate future policy changes, and even contribute their expertise to inform evidence-based regulations.

Finally, sustainability in engineering extends beyond environmental considerations to encompass economic and social sustainability as well. Engineers must strive to create designs that meet present needs without compromising the ability of future generations to meet their own needs. This involves balancing often competing priorities to create solutions that are truly sustainable in the long term.

By integrating these eight critical considerations into every stage of the design process, engineers can create solutions that are not only innovative and functional but also responsible and beneficial to society as a whole. As we move forward into an era of rapid technological advancement and global challenges, the ability to navigate these interconnected factors will be crucial for engineering solutions that truly serve humanity and the planet.

This chapter aims to provide a comprehensive exploration of each of these considerations, equipping aspiring engineers with the knowledge and mindset needed to approach design challenges holistically. By doing so, we can ensure that our innovations not only solve immediate problems but also contribute positively to a sustainable, equitable, and thriving future for all.

18.1 Safety

Safety is paramount in engineering design, serving as the cornerstone of responsible and ethical practice across all disciplines. As engineers, our primary obligation is to ensure that our designs, products, and systems do not pose undue risks to users, workers, the public, or the environment. This section delves deeply into the multifaceted nature of safety considerations in engineering design, exploring methodologies, principles, and best practices that should guide every stage of the design process.

18.1.1 Risk Assessment and Management

18.1.1.1 Hazard Identification

The first step in ensuring safety is identifying potential hazards. This process involves a systematic examination of all aspects of a design to uncover anything that could potentially cause harm. Hazards can be physical (e.g., moving parts, electrical components), chemical (e.g., toxic substances), biological (e.g., pathogens), or even psychological (e.g., stress-inducing interfaces).

Techniques for hazard identification include:
- Failure Mode and Effects Analysis (FMEA)
- Hazard and Operability Study (HAZOP)
- Fault Tree Analysis (FTA)
- What-If Analysis

Engineers should consider not only obvious hazards but also those that might arise from foreseeable misuse or under abnormal operating conditions.

18.1.1.2 Risk Evaluation

Once hazards are identified, the next step is to evaluate the associated risks. This involves assessing both the likelihood of a hazard occurring and the severity of its potential consequences. Risk can be quantified using various methods, including:
- Risk matrices
- Probabilistic risk assessment
- Quantitative risk assessment (QRA)

It's crucial to consider both individual risks and cumulative risks that may arise from the interaction of multiple hazards.

18.1.1.3 Risk Mitigation Strategies

After evaluating risks, engineers must develop and implement appropriate mitigation strategies. The hierarchy of risk control provides a framework for this:
1. Elimination: Remove the hazard entirely.
2. Substitution: Replace the hazard with a less dangerous alternative.
3. Engineering controls: Implement physical changes to isolate people from the hazard.
4. Administrative controls: Change work procedures to reduce exposure to the hazard.

5. Personal protective equipment: Use PPE as a last line of defense.

Engineers should strive to implement controls as high up this hierarchy as possible, as these tend to be more effective and reliable.

18.1.2 Inherently Safe Design

The concept of inherently safe design goes beyond mere risk mitigation to fundamentally rethinking how we approach safety in engineering. This philosophy, pioneered by Trevor Kletz in the chemical industry, advocates for eliminating or minimizing hazards at the source rather than relying on add-on protective measures.

Key principles of inherently safe design include:

18.1.2.1 Minimization

Reduce the amount of hazardous materials or energy in the system. For example, using smaller quantities of chemicals in a production process or lower voltages in electrical systems where possible.

18.1.2.2 Substitution

Replace hazardous materials or processes with less dangerous alternatives. This could involve using water-based solvents instead of organic ones or replacing a chemical process with a mechanical one.

18.1.2.3 Moderation

Operate under less hazardous conditions. This might mean using lower temperatures or pressures in a process, or designing equipment to operate within a safer range of parameters.

18.1.2.4 Simplification

Design systems to be as simple as possible, reducing the potential for errors and making it easier to identify and address potential hazards. This principle often aligns with good design practices in general, as simpler systems tend to be more reliable and easier to maintain.

18.1.3 Redundancy and Fail-Safe Design

Even with the best hazard prevention measures, it's crucial to design systems that can handle failures safely. This is where concepts of redundancy and fail-safe design come into play.

18.1.3.1 Redundancy

Redundancy involves incorporating backup systems or components that can take

over if the primary system fails. This is particularly important in critical systems where failure could have catastrophic consequences. Examples include:
- Multiple engines on aircraft
- Backup power systems in hospitals
- Redundant control systems in nuclear power plants

When implementing redundancy, it's important to consider common mode failures that could affect both the primary and backup systems simultaneously.

18.1.3.2 Fail-Safe Design

Fail-safe design ensures that if a system or component fails, it does so in a way that minimizes harm. This often involves designing systems to default to a safe state when something goes wrong. Examples include:
- Elevator brakes that engage automatically if the cable fails
- Pressure relief valves that open to prevent explosions
- Railway signals that default to 'stop' if they lose power

18.1.4 Human Factors and Safety

While much of safety design focuses on technical aspects, it's crucial to consider the human element. Human error is a significant factor in many accidents, but well-designed systems can help mitigate this risk.

18.1.4.1 Ergonomics and User-Centered Design

Designing with human physical and cognitive capabilities in mind can significantly reduce the risk of accidents. This includes considerations such as:
- Proper positioning of controls and displays
- Clear and intuitive interfaces
- Appropriate physical design to reduce strain and fatigue

18.1.4.2 Mistake-Proofing (Poka-Yoke)

This concept, originating from Japanese manufacturing, involves designing systems to prevent errors or make them immediately obvious. Examples include:
- USB connectors that can only be inserted one way
- Gas pump nozzles that don't fit into diesel car tanks
- Software interfaces that confirm potentially destructive actions

18.1.4.3 Safety Communication

Clear communication of safety information is crucial. This includes:
- Effective warning labels and signs
- Comprehensive but understandable user manuals

- Training programs for complex or potentially dangerous systems

18.1.5 Safety Standards and Regulations

Engineers must be familiar with and adhere to relevant safety standards and regulations in their field. These may include:
- Industry-specific standards (e.g., ISO standards, ASME codes)
- National regulations (e.g., OSHA standards in the US)
- International agreements (e.g., the Montreal Protocol for environmental safety)

While compliance with these standards is often legally required, it should be viewed as a minimum baseline rather than the ultimate goal of safety design.

18.1.6 Safety in the Product Lifecycle

Safety considerations should extend throughout the entire lifecycle of a product or system:

18.1.6.1 Design Phase

This is where the foundation for safety is laid. Conducting thorough risk assessments and implementing inherently safe design principles at this stage can prevent many issues down the line.

18.1.6.2 Manufacturing and Construction

Ensuring safety during the production process is crucial. This includes considering the safety of workers involved in manufacturing or construction, as well as implementing quality control measures to ensure the final product meets safety specifications.

18.1.6.3 Operation and Maintenance

Designs should account for safe operation and maintenance. This includes providing clear instructions, incorporating features that facilitate safe maintenance (e.g., lockout/tagout capabilities), and designing for ease of inspection.

18.1.6.4 Decommissioning and Disposal

The end-of-life stage of a product or system should also be considered from a safety perspective. This might involve designing for easy and safe disassembly, or ensuring that potentially hazardous materials can be disposed of safely.

18.1.7 Emerging Technologies and Safety Challenges

As technology advances, new safety challenges emerge. Engineers must stay informed about these developments and consider their safety implications. Some areas of particular concern include:

18.1.7.1 Artificial Intelligence and Autonomous Systems

As systems become more autonomous, ensuring their safe operation becomes more complex. This includes considerations such as algorithmic bias, the ability to handle unexpected situations, and the ethical implications of decision-making AI.

18.1.7.2 Internet of Things (IoT) and Cybersecurity

With more devices connected to the internet, cybersecurity becomes a crucial safety consideration. A hacked IoT device could potentially cause physical harm, making robust security measures essential.

18.1.7.3 Nanotechnology

As we work with materials at increasingly small scales, new safety considerations arise. The potential health and environmental impacts of nanoparticles, for instance, are still not fully understood.

Safety in engineering design is a complex and ever-evolving field. It requires a proactive, holistic approach that considers technical, human, and systemic factors. By integrating safety considerations throughout the design process, from initial concept to final disposal, engineers can create products and systems that not only function well but also protect users, workers, the public, and the environment from harm.

As we face new technological frontiers and global challenges, the importance of safety in engineering design only grows. It is our responsibility as engineers to remain vigilant, to continually educate ourselves about emerging safety concerns, and to always prioritize safety in our work. By doing so, we fulfill our ethical obligation to society and contribute to a safer, more sustainable world.

18.2 Ergonomics

Ergonomics, also known as human factors engineering, is a critical discipline that focuses on optimizing the interaction between humans and the systems they use. In the context of engineering design, ergonomics plays a crucial role in creating products, workspaces, and systems that are not only functional but also comfortable, efficient, and safe for human use.

The primary goal of ergonomics is to fit the task to the human, rather than forcing the human to adapt to the task. This approach leads to designs that reduce physical and cognitive strain, enhance productivity, and minimize the risk of injuries or errors. As such, ergonomics is an essential consideration in virtually every field of engineering, from consumer product design to industrial systems and software interfaces.

18.2.1 Physical Ergonomics

Physical ergonomics deals with human anatomical, anthropometric, physiological, and biomechanical characteristics as they relate to physical activity. This aspect of ergonomics is crucial in designing products and workspaces that accommodate the human body's capabilities and limitations.

18.2.1.1 Anthropometry and Design

Anthropometry, the study of human body measurements, is fundamental to physical ergonomics. Engineers must consider the variability in human body sizes and proportions when designing products or spaces. This involves:
- Using anthropometric data to determine appropriate dimensions for products or workspaces
- Designing for adjustability to accommodate a wide range of users (e.g., adjustable chairs, steering wheels)
- Considering the needs of specific populations, including children, the elderly, or people with disabilities

18.2.1.2 Biomechanics and Musculoskeletal Considerations

Understanding how the human body moves and the forces it can comfortably exert is crucial for ergonomic design. Key considerations include:
- Designing to minimize awkward postures, repetitive motions, and excessive force requirements
- Optimizing the positioning of controls and displays to reduce physical strain
- Considering the impact of prolonged static postures in designs for workstations or vehicles

18.2.1.3 Environmental Factors

Physical ergonomics also encompasses environmental factors that affect human comfort and performance, such as:
- Lighting design to reduce eye strain and enhance visibility
- Acoustic design to manage noise levels and improve communication
- Thermal comfort considerations in building and vehicle design

18.2.2 Cognitive Ergonomics

Cognitive ergonomics focuses on mental processes, such as perception, memory, reasoning, and motor response, as they affect interactions among humans and other elements of a system. This aspect of ergonomics is increasingly important in our technology-driven world.

18.2.2.1 Information Processing and Interface Design

Key considerations in cognitive ergonomics include:
- Designing interfaces that align with human cognitive processes and limitations
- Optimizing information presentation to reduce mental workload and enhance decision-making
- Considering factors such as attention, perception, and memory in the design of warning systems and instructional materials

18.2.2.2 Human-Computer Interaction

As digital interfaces become ubiquitous, cognitive ergonomics plays a crucial role in their design:
- Creating intuitive navigation systems and menu structures
- Implementing consistent and meaningful icons and symbols
- Designing for different levels of user expertise and accommodating learning curves

18.2.2.3 Mental Workload and Decision Support

In complex systems, cognitive ergonomics helps manage mental workload and support decision-making:
- Designing automation systems that complement human capabilities rather than replacing them entirely
- Creating decision support tools that provide relevant information without overwhelming the user
- Implementing error-prevention and error-recovery mechanisms in critical systems

18.2.3 Organizational Ergonomics

Organizational ergonomics, also known as macroergonomics, is concerned with the optimization of sociotechnical systems, including their organizational structures, policies, and processes.

18.2.3.1 Work System Design

This involves:
- Designing workflows and processes that balance efficiency with human well-being
- Considering the social and cultural aspects of work environments in design decisions
- Implementing participatory ergonomics approaches that involve workers in the design process

18.2.3.2 Communication and Collaboration

Organizational ergonomics also focuses on:
- Designing systems that facilitate effective communication within teams and organizations
- Creating collaborative workspaces that support both individual and group work
- Implementing technologies that enhance remote collaboration while considering ergonomic principles

18.2.4 Universal Design and Accessibility

An important aspect of ergonomics in engineering design is the principle of universal design – creating products and environments that are accessible and usable by people of all ages and abilities, to the greatest extent possible.

18.2.4.1 Inclusive Design Principles

Key considerations include:
- Designing for flexibility in use to accommodate a wide range of individual preferences and abilities
- Ensuring that designs are simple and intuitive to use, regardless of the user's experience, knowledge, language skills, or current concentration level
- Providing perceptible information through different sensory modes (visual, auditory, tactile)

18.2.4.2 Assistive Technologies

Ergonomic design also involves:
- Developing specialized assistive technologies for people with disabilities
- Ensuring compatibility between mainstream products and assistive technologies
- Considering how designs can be adapted or customized for individual needs

18.2.5 Evaluation and Testing in Ergonomic Design

Implementing ergonomic principles in engineering design requires thorough evaluation and testing throughout the design process.

18.2.5.1 Methods of Evaluation

Common evaluation methods include:
- Usability testing with representative users
- Anthropometric analysis and fit testing
- Biomechanical analysis, including digital human modeling
- Cognitive task analysis
- Heuristic evaluations by ergonomics experts

18.2.5.2 Measurement Tools

Various tools and techniques are used in ergonomic evaluation, such as:
- Motion capture systems for analyzing body movements
- Eye-tracking devices for studying visual attention and information processing
- Electromyography (EMG) for measuring muscle activity
- Pressure mapping systems for assessing body pressure distribution

18.2.5.3 Iterative Design Process

Ergonomic design typically involves an iterative process of:
- Initial design based on ergonomic principles and guidelines
- Prototype development
- User testing and evaluation
- Design refinement based on evaluation results

18.2.6 Emerging Trends in Ergonomics

As technology and work environments evolve, new challenges and opportunities arise in the field of ergonomics.

18.2.6.1 Ergonomics in Remote and Flexible Work

With the increase in remote and flexible work arrangements, ergonomics must address:
- Designing for adaptable home office setups
- Developing ergonomic guidelines for mobile device use
- Creating virtual environments that support well-being and productivity

18.2.6.2 Ergonomics in Virtual and Augmented Reality

As VR and AR technologies become more prevalent, ergonomic considerations include:
- Minimizing physical discomfort and motion sickness in VR environments
- Designing intuitive interaction methods for AR interfaces
- Balancing immersion with awareness of the physical environment

18.2.6.3 AI and Adaptive Ergonomics

Emerging technologies are enabling more personalized ergonomic solutions:
- AI-driven systems that adapt to individual user behavior and preferences
- Smart furniture and devices that automatically adjust to optimal ergonomic settings
- Wearable technologies that provide real-time ergonomic feedback and guidance

Ergonomics is a critical consideration in engineering design, encompassing physical, cognitive, and organizational aspects of human-system interaction. By integrating ergonomic principles throughout the design process, engineers can create products, workspaces, and systems that are not only functional but also comfortable, efficient, and safe for human use.

As we continue to push the boundaries of technology and redefine how we work and interact with our environment, the role of ergonomics in engineering design becomes increasingly important. It is essential for engineers to stay informed about ergonomic principles and emerging trends, and to consistently apply this knowledge in their design work. By doing so, we can create designs that truly serve human needs and enhance quality of life.

18.3 Health Considerations

Health considerations in engineering design extend beyond immediate safety concerns to encompass the long-term effects of engineered products, systems, and environments on human wellbeing. As engineers, we have a responsibility to create designs that not only avoid causing harm but actively promote health and wellness. This section explores the multifaceted nature of health considerations in engineering design, covering both physical and mental health aspects, and discussing how engineers can incorporate health-promoting features into their work.

18.3.1 Physical Health Considerations

18.3.1.1 Exposure to Harmful Substances

One of the primary health considerations in engineering design is minimizing exposure to harmful substances. This includes:

- Chemical Hazards: Engineers must consider the toxicity of materials used in products or processes. This involves selecting safer alternatives where possible, implementing effective containment measures, and designing for proper ventilation in industrial settings.
- Particulate Matter: In fields such as civil engineering and manufacturing, controlling exposure to dust, fumes, and other particulates is crucial. This might involve designing effective filtration systems or dust suppression mechanisms.
- Radiation: In medical, nuclear, and some industrial applications, protecting users and workers from harmful radiation is paramount. This includes designing appropriate shielding and implementing fail-safe systems.

18.3.1.2 Ergonomic Health Impacts

While ergonomics is often discussed separately, it has significant implications for

physical health:
- Musculoskeletal Disorders: Designs should minimize the risk of repetitive strain injuries, back problems, and other musculoskeletal issues. This applies to everything from hand tools to office furniture and industrial workstations.
- Cardiovascular Health: Sedentary behavior, often influenced by poor workplace design, is linked to cardiovascular problems. Engineers should consider how their designs can encourage movement and discourage prolonged sitting.

18.3.1.3 Noise and Vibration

Exposure to excessive noise or vibration can lead to various health issues:
- Hearing Loss: In industrial settings, vehicle design, and consumer products, managing noise levels is crucial for preventing hearing damage.
- Vibration-related Disorders: Prolonged exposure to vibration, particularly in industrial equipment and vehicles, can lead to circulatory and neurological problems. Engineers must design to minimize harmful vibrations or provide adequate isolation.

18.3.1.4 Lighting and Visual Health

Proper lighting design is essential for visual health and overall wellbeing:
- Eye Strain: In both physical and digital environments, engineers must consider lighting levels, glare reduction, and contrast to minimize eye strain.
- Circadian Rhythm: Lighting design in buildings and even in digital devices can impact circadian rhythms. Engineers should consider how their designs affect natural sleep-wake cycles.

18.3.2 Mental Health Considerations

18.3.2.1 Stress and Cognitive Load

Engineering designs can significantly impact stress levels and cognitive functioning:
- Information Overload: In software and interface design, engineers must balance providing necessary information with avoiding cognitive overwhelm.
- Environmental Stress: In architectural and urban design, considerations such as noise levels, crowding, and aesthetics can influence stress and mental wellbeing.

18.3.2.2 Social Interaction and Isolation

The way we design spaces and systems can profoundly affect social interaction:
- Collaborative Spaces: In workplace and urban design, creating opportunities for positive social interaction can promote mental health.

- Digital Design: In an increasingly digital world, engineers must consider how their designs affect real-world social connections and the potential for social isolation.

18.3.2.3 Autonomy and Control

Designs that allow users a sense of control over their environment can positively impact mental health:
- Customization: Allowing users to customize their experience, whether in physical spaces or digital interfaces, can enhance wellbeing.
- Predictability and Consistency: Designs that behave in predictable ways can reduce stress and anxiety, particularly in complex systems.

18.3.3 Health-Promoting Design

Beyond mitigating negative health impacts, engineers can actively design for positive health outcomes:

18.3.3.1 Physical Activity Promotion

- Active Design in Architecture: Incorporating features like prominent, attractive staircases can encourage physical activity in buildings.
- Urban Planning: Designing walkable cities with good cycling infrastructure can promote regular physical activity.
- Product Design: Creating engaging fitness equipment or health-tracking devices can motivate increased physical activity.

18.3.3.2 Healthy Eating Promotion

- Food Technology: Developing technologies for healthier food processing or preservation.
- Kitchen Design: Creating kitchens (both domestic and commercial) that make healthy food preparation easier and more appealing.

18.3.3.3 Mental Wellbeing Promotion

- Biophilic Design: Incorporating natural elements into built environments to reduce stress and improve wellbeing.
- Mindfulness Technology: Developing apps or devices that encourage mindfulness and stress reduction practices.

18.3.3.4 Sleep Promotion

- Lighting Systems: Designing lighting that adjusts throughout the day to support natural circadian rhythms.

- Bedroom Design: Creating sleep-promoting environments through consideration of factors like temperature control, sound insulation, and air quality.

18.3.4 Inclusive Health Design

Health considerations in engineering must account for diverse populations:

18.3.4.1 Accessibility

- Designing for Disabilities: Ensuring that health-promoting features are accessible to people with various physical and cognitive abilities.
- Age-Friendly Design: Considering the health needs of both young and aging populations in design decisions.

18.3.4.2 Cultural Considerations

- Cultural Sensitivity: Recognizing that health practices and priorities may vary across cultures and designing flexibly to accommodate these differences.
- Global Health: Considering how designs might impact health in different global contexts, particularly in resource-limited settings.

18.3.5 Health Monitoring and Feedback Systems

Advancements in technology allow for the integration of health monitoring into various designs:

18.3.5.1 Wearable Technology

- Health Tracking: Designing wearable devices that can monitor vital signs, activity levels, and other health indicators.
- Feedback Systems: Creating intuitive ways to provide health-related feedback to users, encouraging healthy behaviors.

18.3.5.2 Smart Environments

- Environmental Monitoring: Designing systems that monitor air quality, temperature, and other environmental factors that impact health.
- Adaptive Environments: Creating spaces that automatically adjust based on health-related data (e.g., lighting that adapts to circadian rhythms).

18.3.6 Evaluation and Testing for Health Impacts

Assessing the health impacts of designs is crucial:

18.3.6.1 Health Impact Assessments

- Systematic Evaluation: Conducting thorough assessments of how designs might

impact various aspects of health, both positively and negatively.
- Long-term Studies: Where possible, engaging in or considering longitudinal studies to understand the long-term health effects of designs.

18.3.6.2 User Testing

- Diverse User Groups: Ensuring that health-related features are tested with a diverse range of users to understand varied health impacts.
- Real-World Testing: Moving beyond laboratory settings to understand how designs impact health in real-life contexts.

18.3.7 Ethical Considerations in Health-Related Design

18.3.7.1 Privacy and Data Security

- Health Data Protection: Ensuring that any health data collected through designed systems is securely stored and transmitted.
- Informed Consent: Designing systems that clearly communicate how health data will be used and allow users to make informed choices.

18.3.7.2 Unintended Consequences

- Holistic Impact Assessment: Considering how health-focused designs might have unintended negative impacts on other aspects of wellbeing or society.
- Adaptive Design: Creating systems that can be adjusted or updated as new health information becomes available.

18.3.8 Interdisciplinary Collaboration

Effective health consideration in engineering design often requires collaboration:

18.3.8.1 Working with Health Professionals

- Medical Expertise: Engaging with healthcare professionals to understand specific health impacts and opportunities.
- Public Health Collaboration: Working with public health experts to understand broader health trends and priorities.

18.3.8.2 Psychological and Social Science Input

- Behavioral Insights: Collaborating with psychologists to understand how designs can effectively promote healthy behaviors.
- Sociological Perspectives: Considering the social determinants of health and how designs interact with these factors.

Health considerations in engineering design are complex and far-reaching, encompassing both the prevention of negative health impacts and the active promotion of wellbeing. As our understanding of health evolves and new technologies emerge, the opportunities for health-promoting design continue to expand.

Engineers have a unique opportunity – and responsibility – to create designs that contribute positively to human health. This requires a holistic approach that considers physical and mental health, accounts for diverse populations, leverages new technologies, and engages in rigorous evaluation.

By prioritizing health considerations throughout the design process, engineers can create products, systems, and environments that not only serve their primary functions but also enhance quality of life and contribute to a healthier society. As we face global health challenges and rapidly changing technological landscapes, the integration of health considerations into engineering design becomes increasingly crucial for creating a sustainable and thriving future.

18.4 Environmental Considerations

Environmental considerations have become increasingly crucial in engineering design as we face global challenges such as climate change, resource depletion, and ecosystem degradation. Engineers play a pivotal role in shaping the world around us, and with this power comes the responsibility to design with environmental sustainability in mind. This section explores the multifaceted nature of environmental considerations in engineering design, covering topics from lifecycle assessment to biomimicry, and discussing strategies for minimizing environmental impact while maximizing positive contributions to ecosystem health.

18.4.1 Life Cycle Assessment (LCA)

18.4.1.1 Understanding LCA

Life Cycle Assessment is a comprehensive approach to evaluating the environmental impacts of a product or system throughout its entire life cycle, from raw material extraction to disposal or recycling. Key stages typically include:
- Raw material extraction and processing
- Manufacturing and production
- Distribution and transportation
- Use and maintenance
- End-of-life (disposal, recycling, or reuse)

18.4.1.2 Implementing LCA in Design

Engineers should consider LCA early in the design process to:

- Identify hotspots of environmental impact
- Compare different design alternatives
- Inform material selection and manufacturing processes
- Guide decisions on energy efficiency and resource use

18.4.1.3   LCA Tools and Methodologies

Various tools and methodologies exist to conduct LCA, including:
- Software tools like SimaPro, GaBi, or OpenLCA
- Standardized methodologies such as ISO 14040 and 14044
- Simplified LCA approaches for quick assessments in early design stages

18.4.2   Circular Economy Design

18.4.2.1   Principles of Circular Economy

The circular economy model aims to eliminate waste and maximize resource efficiency. Key principles include:
- Designing out waste and pollution
- Keeping products and materials in use
- Regenerating natural systems

18.4.2.2   Strategies for Circular Design

Engineers can incorporate circular economy principles through:
- Design for durability and longevity
- Modularity and ease of repair
- Design for disassembly and recycling
- Use of recycled or renewable materials
- Implementation of take-back systems or product-as-a-service models

18.4.2.3   3.3 Challenges and Opportunities

While circular design presents challenges such as complex supply chains and changing business models, it also offers opportunities for innovation and new value creation.

18.4.3   Energy Efficiency and Renewable Energy Integration

18.4.3.1   Energy Efficiency in Product Design

Considerations for energy-efficient design include:
- Optimizing power consumption in electronic devices
- Improving thermal insulation in buildings

- Enhancing engine efficiency in vehicles
- Implementing smart power management systems

18.4.3.2 Renewable Energy Integration

Engineers should consider how to integrate renewable energy sources into their designs:
- Incorporating solar panels into building design
- Designing products to be compatible with renewable energy sources
- Developing energy storage solutions for intermittent renewable sources
- Creating smart grids and energy management systems

18.4.3.3 Embodied Energy

It's crucial to consider not just operational energy use, but also the embodied energy in materials and manufacturing processes.

18.4.4 Water Conservation and Management

18.4.4.1 Water-Efficient Design

Strategies for water conservation in design include:
- Developing low-flow and water-saving fixtures
- Designing efficient irrigation systems
- Implementing water recycling and reuse systems in industrial processes
- Creating drought-resistant landscaping designs

18.4.4.2 Stormwater Management

Engineers should consider stormwater management in urban and infrastructure design:
- Designing permeable pavements and green roofs
- Creating bioswales and rain gardens
- Implementing urban water retention and slow-release systems

18.4.4.3 Water Quality Protection

Designs should aim to protect water quality by:
- Minimizing pollutant runoff from industrial and agricultural systems
- Developing effective water treatment technologies
- Creating buffer zones to protect water bodies in land development projects

18.4.5 Material Selection and Resource Efficiency

18.4.5.1 Sustainable Materials

Engineers should prioritize materials that have lower environmental impacts:
- Renewable materials (e.g., sustainably sourced wood, bioplastics)
- Recycled and recyclable materials
- Materials with low embodied energy and carbon footprint
- Non-toxic and biodegradable materials

18.4.5.2 Resource Efficiency

Strategies for resource efficiency include:
- Lightweighting to reduce material use
- Optimizing manufacturing processes to minimize waste
- Implementing just-in-time production to reduce inventory waste
- Designing for material recovery and recycling

18.4.5.3 Critical Materials and Supply Chain Considerations

Engineers should be aware of the environmental and social impacts of material sourcing, particularly for rare earth elements and other critical materials.

18.4.6 Biodiversity and Ecosystem Protection

18.4.6.1 Habitat Preservation and Restoration

Engineering projects, especially in civil and environmental engineering, should consider:
- Minimizing habitat disruption in construction projects
- Incorporating wildlife corridors in infrastructure design
- Restoring degraded ecosystems as part of project implementation

18.4.6.2 Biomimicry and Nature-Inspired Design

Engineers can look to nature for sustainable design solutions:
- Emulating natural forms and processes for efficiency (e.g., wind turbine blades inspired by whale fins)
- Developing materials and structures inspired by natural systems
- Creating self-healing or adaptive systems based on biological principles

18.4.6.3 Urban Ecology

In urban design and architecture, consider:
- Creating green spaces and urban forests

18.4 Environmental Considerations

- Designing for urban biodiversity (e.g., green roofs, pollinator-friendly landscapes)
- Implementing urban farming and food production systems

18.4.7 Pollution Prevention and Control

18.4.7.1 Air Quality

Engineers should design to minimize air pollution through:
- Developing clean energy technologies
- Improving filtration and emissions control systems
- Designing for reduced volatile organic compound (VOC) emissions in products and buildings

18.4.7.2 Noise Pollution

Consider noise reduction in:
- Vehicle and machinery design
- Urban planning and building design
- Industrial process engineering

18.4.7.3 Light Pollution

Mitigate light pollution through:
- Efficient and directed outdoor lighting design
- Use of adaptive lighting systems
- Consideration of impacts on wildlife and human health in lighting plans

18.4.8 Climate Change Mitigation and Adaptation

18.4.8.1 Carbon Footprint Reduction

Engineers should prioritize low-carbon designs:
- Maximizing energy efficiency
- Incorporating renewable energy
- Using low-carbon materials
- Designing for reduced emissions throughout the product lifecycle

18.4.8.2 Climate Resilience

Design for resilience to climate change impacts:
- Creating flood-resistant infrastructure
- Developing drought-tolerant agricultural systems
- Designing buildings for extreme weather events

- Implementing adaptive management systems in urban planning

18.4.9 Waste Reduction and Management

18.4.9.1 Design for Reduced Waste

Strategies include:
- Minimizing packaging
- Creating durable, long-lasting products
- Designing for easy repair and upgrade
- Implementing modular design for component replacement

18.4.9.2 Waste-to-Resource Systems

Consider how waste from one process can become a resource for another:
- Designing industrial symbiosis systems
- Creating energy-from-waste technologies
- Developing upcycling processes for waste materials

18.4.10 Environmental Impact Assessment and Monitoring

18.4.10.1 Predictive Modeling

Use environmental modeling to:
- Predict potential impacts of designs on ecosystems
- Simulate long-term environmental effects
- Optimize designs for minimal environmental impact

18.4.10.2 Monitoring Systems

Incorporate environmental monitoring into designs:
- Developing sensors and IoT systems for real-time environmental data collection
- Creating adaptive systems that respond to environmental conditions
- Implementing predictive maintenance to prevent environmental incidents

Environmental considerations in engineering design are no longer optional; they are fundamental to responsible and sustainable engineering practice. As we face unprecedented environmental challenges, engineers have the opportunity and responsibility to create designs that not only minimize negative impacts but actively contribute to environmental restoration and sustainability.

This requires a holistic approach that considers the entire lifecycle of products and systems, embraces circular economy principles, prioritizes energy and resource efficiency, protects biodiversity, and addresses climate change. It also demands interdisciplinary collaboration, as environmental issues are complex and interconnected.

By integrating these environmental considerations throughout the design process, engineers can drive innovation, create more resilient and efficient systems, and contribute to a more sustainable future. As technology advances and our understanding of environmental systems deepens, the potential for environmentally positive design continues to grow.

The challenge for engineers is to balance these environmental considerations with other design requirements, including functionality, cost, and user needs. However, by viewing environmental sustainability not as a constraint but as a driver of innovation, engineers can create designs that are not only environmentally sound but also economically viable and socially beneficial.

In the face of global environmental challenges, the role of environmentally conscious engineering design has never been more critical. It is through thoughtful, innovative, and responsible design that we can hope to create a sustainable and thriving world for current and future generations.

18.5 Social Considerations

Social considerations in engineering design encompass the ways in which engineered products, systems, and environments impact and interact with society. As engineers, we must recognize that our designs do not exist in isolation but are intrinsically linked to the social fabric of communities and cultures. This section explores the multifaceted nature of social considerations in engineering design, discussing how engineers can create solutions that are not only technically sound but also socially responsible and beneficial.

18.5.1 Understanding Social Impact

18.5.1.1 Social Impact Assessment

Social Impact Assessment (SIA) is a crucial tool for engineers to evaluate the potential social consequences of their designs. Key aspects include:
- Identifying stakeholders affected by the design
- Assessing both positive and negative social impacts
- Considering short-term and long-term social effects
- Evaluating impacts on different social groups (e.g., by age, gender, socioeconomic status)

18.5.1.2 Types of Social Impacts

Engineers should consider various types of social impacts, including:

- Economic impacts (e.g., job creation, income distribution)
- Cultural impacts (e.g., effects on local traditions and ways of life)
- Demographic impacts (e.g., population movements, changes in community composition)
- Health and well-being impacts
- Effects on social relationships and community structures

18.5.2 Participatory Design and Community Engagement

18.5.2.1 Importance of Participatory Design

Involving communities and end-users in the design process is crucial for creating socially responsible solutions. Benefits include:
- Ensuring designs meet actual community needs
- Incorporating local knowledge and perspectives
- Building community support and ownership
- Identifying potential social issues early in the design process

18.5.2.2 Methods of Community Engagement

Engineers can employ various methods to engage communities:
- Public consultations and town hall meetings
- Participatory workshops and design charrettes
- Surveys and interviews with community members
- Collaboration with local organizations and leaders
- Prototyping and user testing with community involvement

18.5.3 Cultural Sensitivity and Diversity

18.5.3.1 Cultural Considerations in Design

Engineers must be aware of and respect cultural differences when designing. This involves:
- Understanding local customs, values, and beliefs
- Considering how designs might impact or be perceived by different cultural groups
- Avoiding cultural appropriation or insensitivity in design choices

18.5.3.2 Designing for Diversity

Inclusive design that considers diverse populations is essential:
- Addressing the needs of various age groups, genders, and abilities
- Considering linguistic diversity in user interfaces and instructions

- Designing for different family structures and living arrangements
- Accounting for diverse religious and cultural practices

18.5.4 Accessibility and Universal Design

18.5.4.1 Principles of Universal Design

Universal design aims to create solutions usable by all people, to the greatest extent possible, without the need for adaptation. Key principles include:
- Equitable use
- Flexibility in use
- Simple and intuitive use
- Perceptible information
- Tolerance for error
- Low physical effort
- Size and space for approach and use

18.5.4.2 Accessibility Considerations

Engineers should design with accessibility in mind, considering:
- Physical accessibility for people with mobility impairments
- Visual and auditory accessibility for sensory impairments
- Cognitive accessibility for neurological diversity
- Digital accessibility in software and web design

18.5.5 Social Equity and Justice

18.5.5.1 Addressing Social Inequalities

Engineers have a responsibility to consider how their designs might impact social equity:
- Avoiding designs that exacerbate existing social disparities
- Creating solutions that promote equal access to resources and opportunities
- Considering affordability and access for low-income communities

18.5.5.2 Environmental Justice

Environmental justice concerns should be integrated into design decisions:
- Avoiding disproportionate environmental burdens on marginalized communities
- Ensuring equitable distribution of environmental benefits
- Involving affected communities in environmental decision-making processes

18.5.6 Technology and Social Change

18.5.6.1 Societal Impacts of Technological Innovation

Engineers must consider how their innovations might drive social change:
- Impacts on employment and job markets
- Changes in social interactions and communication patterns
- Effects on privacy and data security
- Potential for technology addiction or overreliance

18.5.6.2 Ethical Considerations in Emerging Technologies

As new technologies emerge, engineers face complex ethical considerations:
- Addressing bias in artificial intelligence and machine learning systems
- Considering the societal implications of genetic engineering and biotechnology
- Evaluating the social impacts of automation and robotics
- Addressing privacy concerns in IoT and smart technologies

18.5.7 Social Sustainability

18.5.7.1 Long-term Social Viability

Engineers should design for long-term social sustainability:
- Creating solutions that can adapt to changing social needs
- Considering the intergenerational impacts of designs
- Promoting social resilience in the face of challenges like climate change

18.5.7.2 Social Capital and Community Building

Designs can contribute to building social capital:
- Creating public spaces that foster community interaction
- Developing technologies that facilitate social connections
- Designing infrastructure that supports community activities and events

18.5.8 Education and Capacity Building

18.5.8.1 Designing for Learning and Skill Development

Engineers can contribute to social development through designs that:
- Support education and lifelong learning
- Facilitate skill development and capacity building
- Promote technology literacy and digital inclusion

18.5 Social Considerations

18.5.8.2 Knowledge Transfer

Consider how designs can facilitate knowledge transfer:
- Creating user-friendly interfaces and documentation
- Designing for ease of maintenance and repair by local communities
- Developing training programs as part of technology implementation

18.5.9 Health and Well-being

18.5.9.1 Public Health Considerations

Engineers should consider how their designs impact public health:
- Creating environments that promote physical activity
- Designing for improved air and water quality
- Developing technologies for healthcare access and delivery

18.5.9.2 Mental Health and Social Well-being

Consider the psychological and social impacts of designs:
- Creating spaces that reduce stress and promote mental well-being
- Designing technologies that support work-life balance
- Considering the impacts of design on social connections and community cohesion

Social considerations in engineering design are fundamental to creating solutions that truly serve humanity and contribute positively to society. By integrating these considerations throughout the design process, engineers can create more holistic, equitable, and sustainable solutions.

This approach requires engineers to think beyond technical specifications and consider the broader social context in which their designs will exist. It involves engaging with communities, understanding diverse needs and perspectives, and anticipating both the intended and unintended social consequences of our designs.

As our world becomes increasingly interconnected and complex, the importance of socially responsible engineering design only grows. Engineers have the power to shape not just physical structures and technologies, but also social structures and human experiences. With this power comes the responsibility to create designs that promote social equity, cultural respect, accessibility, and overall well-being.

By embracing social considerations, engineers can drive positive social change, create more resilient and inclusive communities, and contribute to a more just and sustainable world. This holistic approach to engineering not only leads to better solutions but also enhances the profession's ability to address global challenges and improve quality of life for all.

As we move forward, it is crucial for engineering education and practice to continue emphasizing the importance of social considerations. By doing so, we can ensure that future generations of engineers are equipped to create designs that are not only technically excellent but also socially beneficial and ethically sound.

18.6 Ethical Considerations

Ethics form the moral backbone of engineering practice, guiding decisions and actions throughout the design process. As engineers, we are tasked not only with creating functional and innovative solutions but also with ensuring that our designs align with ethical principles and contribute positively to society. This section explores the multifaceted nature of ethical considerations in engineering design, discussing key principles, common ethical dilemmas, and strategies for navigating complex ethical landscapes.

18.6.1 Fundamental Ethical Principles in Engineering

18.6.1.1 Professional Codes of Ethics

Most engineering societies have established codes of ethics that provide a framework for ethical decision-making. Common principles include:
- Holding paramount the safety, health, and welfare of the public
- Performing services only in areas of competence
- Issuing public statements only in an objective and truthful manner
- Acting as faithful agents or trustees for each employer or client
- Avoiding deceptive acts and conflicts of interest
- Conducting oneself honorably, responsibly, ethically, and lawfully

18.6.1.2 Ethical Theories Relevant to Engineering

Understanding broader ethical theories can provide a foundation for decision-making:
- Utilitarianism: Considering the greatest good for the greatest number
- Deontological ethics: Focusing on the inherent rightness or wrongness of actions
- Virtue ethics: Emphasizing the moral character of the decision-maker
- Care ethics: Prioritizing relationships and responsibilities to others

18.6.2 Safety and Risk

18.6.2.1 Prioritizing Public Safety

The paramount ethical consideration in engineering design is ensuring public

safety. This involves:
- Conducting thorough risk assessments
- Implementing appropriate safety factors
- Considering potential misuse or unintended consequences of designs
- Prioritizing safety even when it conflicts with other design goals or stakeholder interests

18.6.2.2 Ethical Dimensions of Risk Communication

Engineers have an ethical obligation to:
- Clearly communicate risks associated with their designs to stakeholders
- Avoid downplaying or obscuring potential hazards
- Provide accurate information to enable informed decision-making

18.6.2.3 Balancing Innovation and Caution

Engineers must navigate the ethical tension between:
- Pushing technological boundaries to create innovative solutions
- Exercising appropriate caution to prevent harm

18.6.3 Environmental Ethics

18.6.3.1 Sustainability and Intergenerational Justice

Ethical engineering design considers long-term environmental impacts:
- Minimizing resource depletion and environmental degradation
- Considering the rights and needs of future generations
- Designing for circular economy principles and waste reduction

18.6.3.2 Biodiversity and Ecosystem Health

Engineers have an ethical responsibility to:
- Minimize harm to ecosystems and biodiversity
- Consider the intrinsic value of nature, not just its utility to humans
- Incorporate biomimicry and nature-inspired design where appropriate

18.6.3.3 Climate Change Ethics

The ethical implications of climate change for engineers include:
- Prioritizing low-carbon and climate-resilient designs
- Considering global equity in climate change mitigation and adaptation
- Transparently communicating the climate impacts of designs

18.6.4 Social Justice and Equity

18.6.4.1 Accessibility and Inclusive Design

Ethical engineering prioritizes inclusivity:
- Designing for diverse user groups, including people with disabilities
- Considering how designs might exclude or disadvantage certain populations
- Promoting universal design principles

18.6.4.2 Global Ethics and Cultural Sensitivity

In an interconnected world, engineers must consider:
- The global impacts of their designs, including on developing countries
- Cultural differences in values, norms, and practices
- Avoiding cultural imperialism in design solutions

18.6.4.3 Addressing Social Inequalities

Ethical engineering design involves:
- Considering how designs might exacerbate or alleviate social inequalities
- Prioritizing solutions that promote social equity and justice
- Engaging with marginalized communities in the design process

18.6.5 Privacy and Data Ethics

18.6.5.1 Data Collection and Use

As technology becomes increasingly data-driven, engineers must consider:
- The ethical implications of data collection and surveillance
- Ensuring informed consent in data collection
- Protecting user privacy and data security

18.6.5.2 Algorithmic Fairness and Transparency

In designing AI and algorithmic systems, ethical considerations include:
- Addressing and mitigating algorithmic bias
- Ensuring transparency and explainability in AI decision-making
- Considering the societal impacts of automation and AI

18.6.5.3 Digital Rights and Internet Ethics

Engineers involved in digital technologies must consider:
- The right to internet access and digital inclusion
- Freedom of expression and censorship issues
- The ethical implications of digital platform design on public discourse

18.6.6 Professional Integrity and Whistleblowing

18.6.6.1 Conflict of Interest

Engineers must navigate potential conflicts between:
- Professional judgment and personal or corporate interests
- Short-term gains and long-term ethical considerations

18.6.6.2 Intellectual Honesty

Ethical engineering practice demands:
- Accurate reporting of data and results
- Acknowledgment of limitations and uncertainties in designs
- Respect for intellectual property rights

18.6.6.3 Ethical Whistleblowing

Engineers may face situations requiring them to:
- Speak out against unethical practices
- Navigate the tension between organizational loyalty and public safety
- Understand legal protections and ethical obligations in whistleblowing

18.6.7 Emerging Technologies and Future Ethics

18.6.7.1 Anticipatory Ethics

As technology advances rapidly, engineers must engage in:
- Proactive consideration of potential ethical issues in emerging technologies
- Scenario planning for various ethical outcomes
- Developing flexible ethical frameworks that can adapt to new challenges

18.6.7.2 Human Enhancement Technologies

Ethical considerations in fields like biotechnology and neurotechnology include:
- The boundaries of human enhancement and identity
- Equitable access to enhancement technologies
- Preserving human autonomy and dignity

18.6.7.3 Artificial General Intelligence (AGI) and Existential Risk

Long-term ethical considerations include:
- The potential impacts of AGI on humanity
- Ensuring that advanced AI systems align with human values
- Balancing technological progress with existential risk mitigation

18.6.8 Ethical Decision-Making Frameworks

18.6.8.1 Structured Ethical Analysis

Engineers can use frameworks like:
- The Seven-Step Guide to Ethical Decision-Making
- The Ethical Matrix
- Value-Sensitive Design methodologies

18.6.8.2 Stakeholder Analysis

Ethical design involves:
- Identifying all stakeholders affected by a design
- Considering the diverse values and interests of different stakeholders
- Balancing competing ethical claims

18.6.8.3 Ethical Impact Assessment

Similar to environmental impact assessments, engineers can conduct:
- Systematic evaluations of the ethical implications of designs
- Scenario analysis of potential ethical outcomes
- Ongoing ethical monitoring throughout the design lifecycle

18.6.9 Ethics Education and Professional Development

18.6.9.1 Integrating Ethics in Engineering Education

To prepare future engineers for ethical challenges:
- Incorporate ethics throughout the engineering curriculum
- Use case studies and real-world ethical dilemmas in teaching
- Develop students' capacities for ethical reasoning and moral imagination

18.6.9.2 Continuing Ethical Education

Practicing engineers should engage in:
- Ongoing professional development in engineering ethics
- Staying informed about emerging ethical issues in their field
- Participating in ethical discussions and debates within the profession

Ethical considerations are integral to responsible and effective engineering design. As engineers, we have a profound responsibility to consider the ethical implications of our work and to strive for designs that not only meet technical specifications but also contribute positively to society and the environment.

Navigating ethical considerations in engineering design often involves balancing competing values, interests, and responsibilities. It requires a combination of ethical knowledge, critical thinking skills, moral courage, and a commitment to the greater good. By integrating ethical considerations throughout the design process, engineers can create more holistic, responsible, and beneficial solutions.

As technology continues to advance and our world faces increasingly complex challenges, the importance of ethical engineering design only grows. From addressing climate change to ensuring the responsible development of AI, engineers play a crucial role in shaping the future of our society and planet.

To meet these challenges, it's essential that the engineering profession continues to emphasize and evolve its approach to ethics. This involves not only adhering to established ethical codes but also actively engaging in ethical deliberation, fostering a culture of ethical awareness, and being willing to speak up and act when faced with ethical dilemmas.

By embracing our ethical responsibilities, we as engineers can enhance the positive impact of our work, maintain public trust in the profession, and contribute to creating a more just, sustainable, and flourishing world for all. The integration of ethical considerations in engineering design is not just a professional obligation—it's an opportunity to leverage our skills and knowledge for the betterment of humanity and our planet.

18.7 Political Considerations

Engineering design, often perceived as a purely technical endeavor, is inextricably linked to the political landscape in which it operates. Political considerations in engineering design encompass a wide range of factors, from regulatory compliance and public policy to geopolitical dynamics and social movements. As engineers, understanding and navigating these political dimensions is crucial for creating designs that are not only technically sound but also socially acceptable, politically viable, and ethically responsible.

This essay explores the multifaceted nature of political considerations in engineering design, discussing key areas where politics intersects with engineering practice, the challenges this intersection presents, and strategies for effectively navigating this complex landscape.

18.7.1 Regulatory Compliance and Policy Frameworks

One of the most direct ways in which politics influences engineering design is through regulations and policy frameworks. Governments at various levels create laws, standards, and guidelines that engineers must adhere to in their designs. These regulations often reflect political priorities and societal values, such as safety, environmental protection, or accessibility.

Key considerations in this area include:

1. Staying informed about current and upcoming regulations: Engineers must continuously update their knowledge of relevant laws and standards, which can change with shifts in political leadership or public sentiment.
2. Anticipating regulatory changes: Forward-thinking engineers consider not just current regulations but also potential future changes, designing with flexibility to adapt to evolving political landscapes.
3. Engaging in the regulatory process: Engineers can contribute their technical expertise to inform policy-making, participating in public consultations or advisory committees.
4. Navigating international regulatory differences: In an increasingly globalized world, engineers often need to design products or systems that comply with regulations across multiple jurisdictions, each influenced by its own political context.

18.7.2 Public Infrastructure and Urban Planning

Engineering projects in public infrastructure and urban planning are deeply intertwined with political processes. These projects often involve significant public funding, impact large populations, and can shape the development of communities for decades.

Political considerations in this domain include:
1. Stakeholder engagement: Engineers must navigate the interests of various stakeholders, including local communities, businesses, and different levels of government.
2. Public-private partnerships: Understanding the political dynamics of collaborations between government entities and private companies is crucial for many large-scale projects.
3. Environmental impact and sustainability: Political pressure around climate change and environmental protection increasingly influences infrastructure design decisions.
4. Social equity: Engineers must consider how their designs impact different communities, addressing concerns about gentrification, accessibility, and fair distribution of resources.

18.7.3 Technology Policy and Innovation

The development and implementation of new technologies are heavily influenced by political factors. Government policies can either foster or hinder technological innovation, and political debates often shape public acceptance of new technologies.

Key areas of consideration include:
1. Research funding priorities: Government decisions on research funding can significantly impact which technologies are developed and brought to market.

2. Intellectual property laws: Understanding and navigating patent systems and other IP protections, which vary across countries, is crucial for many engineering projects.
3. Data privacy and security: Political debates around data protection and cybersecurity directly impact the design of digital systems and products.
4. Ethical guidelines for emerging technologies: Engineers must navigate evolving political discussions around the ethical implications of technologies like AI, genetic engineering, or autonomous systems.

18.7.4 Geopolitical Considerations

In an interconnected world, geopolitical factors can significantly impact engineering design decisions, especially for multinational projects or global supply chains.

Important considerations include:
1. Trade policies and tariffs: Engineers must consider how changing trade relationships between countries might affect material sourcing, manufacturing, and distribution.
2. Technology transfer restrictions: Navigating export control laws and restrictions on sharing sensitive technologies across borders is crucial in many high-tech fields.
3. Global standards harmonization: Engineers often need to reconcile different technical standards across countries, which can be influenced by political relationships and trade agreements.
4. Resource scarcity and geopolitical tensions: The availability and control of critical resources (e.g., rare earth elements) can be subject to geopolitical tensions, impacting design choices and material selection.

18.7.5 Social Movements and Public Opinion

Engineering designs do not exist in a vacuum but are subject to public scrutiny and can be influenced by social movements and shifting public opinions.

Key considerations include:
1. Environmental movements: Growing public concern about climate change and environmental degradation has pushed engineers to prioritize sustainable design practices.
2. Social justice movements: Increased awareness of social inequalities has led to greater emphasis on inclusive and equitable design in various engineering fields.
3. Consumer activism: Engineers must consider how public opinion and consumer boycotts might impact the acceptance and success of their designs.
4. Transparency and accountability: There's growing public demand for transparency in how products are designed, manufactured, and how they impact

society and the environment.

18.7.6 Strategies for Navigating Political Considerations

To effectively navigate these political considerations, engineers can employ several strategies:
1. Develop political literacy: Engineers should strive to understand the political systems and processes that impact their work, going beyond technical knowledge to grasp policy-making processes and stakeholder dynamics.
2. Engage in interdisciplinary collaboration: Working with professionals from fields like public policy, law, and social sciences can help engineers better understand and address political dimensions of their designs.
3. Practice stakeholder engagement: Actively involving and communicating with various stakeholders throughout the design process can help anticipate and address political challenges.
4. Embrace adaptive design approaches: Designing with flexibility to adapt to changing political landscapes can help ensure the long-term viability of engineering solutions.
5. Participate in professional organizations: Engineering societies often engage in policy advocacy and can provide resources for navigating political considerations in the field.

Political considerations are an integral part of the engineering design process, influencing everything from regulatory compliance to public acceptance of new technologies. As engineers, we must recognize that our work does not occur in a political vacuum but is shaped by and contributes to broader societal and political dynamics.

By actively engaging with these political dimensions, engineers can create designs that are not only technically sound but also socially responsible, politically viable, and capable of addressing complex global challenges. This requires expanding our skill set beyond traditional technical expertise to include political literacy, stakeholder engagement, and an understanding of broader societal impacts.

As we face increasingly complex challenges like climate change, rapid technological advancement, and global inequalities, the ability to navigate political considerations in engineering design becomes ever more crucial. By embracing this broader perspective, engineers can enhance their positive impact on society and play a vital role in shaping a more sustainable, equitable, and prosperous future for all.

18.8 Sustainability

Sustainability has emerged as one of the most critical considerations in modern engineering design. As we face unprecedented global challenges such as climate change, resource depletion, and ecosystem degradation, the role of engineers in creating sustainable solutions has never been more crucial. Sustainable engineering design goes beyond mere environmental protection; it encompasses a holistic approach that balances environmental stewardship, economic viability, and social equity. This essay explores the concept of sustainability in engineering design, its key principles, challenges, and strategies for implementation.

18.8.1 Defining Sustainability in Engineering Design

Sustainability in engineering design can be defined as the creation of products, processes, and systems that meet current needs without compromising the ability of future generations to meet their own needs. This definition, inspired by the Brundtland Commission's concept of sustainable development, emphasizes three key pillars:

- **Environmental Sustainability** -- Minimizing negative environmental impacts, conserving resources, and protecting ecosystems.
- **Economic Sustainability** -- Ensuring long-term economic viability and efficiency of designs.
- **Social Sustainability** -- Considering social equity, community well-being, and quality of life in design decisions.

18.8.2 Key Principles of Sustainable Engineering Design

- **Life Cycle Thinking** -- Sustainable design considers the entire life cycle of a product or system, from raw material extraction to end-of-life disposal or recycling. This approach, often implemented through Life Cycle Assessment (LCA), helps identify and mitigate environmental impacts at every stage.
- **Resource Efficiency** -- Maximizing the efficiency of resource use is crucial. This includes:
  - Material efficiency: Using fewer materials, choosing sustainable materials, and designing for recycling or upcycling.
  - Energy efficiency: Minimizing energy consumption during production and use phases.
  - Water efficiency: Reducing water usage and implementing water recycling systems.
- **Waste Minimization** -- Adopting a "cradle-to-cradle" approach rather than a "cradle-to-grave" mentality. This involves designing products that can be fully recycled or biodegraded at the end of their life.
- **Pollution Prevention** -- Designing to minimize or eliminate pollution at the source, rather than relying on end-of-pipe solutions.
- **Ecosystem Preservation** -- Considering the impact of designs on biodiversity and ecosystem services, and striving to preserve or enhance natural habitats.
- **Social Responsibility** -- Ensuring that designs contribute positively to society,

considering factors such as fair labor practices, community impact, and cultural preservation.

18.8.3 Strategies for Implementing Sustainability in Engineering Design

- **Circular Economy Design** -- Embracing circular economy principles in design involves creating products and systems that keep resources in use for as long as possible. Strategies include:
    o Designing for durability and longevity
    o Creating modular designs for easy repair and upgrade
    o Implementing take-back systems and product-as-a-service models
- **Biomimicry** -- Looking to nature for sustainable design solutions. Natural systems have evolved efficient, sustainable solutions over millions of years, and biomimicry seeks to emulate these in engineering design.
- **Green Chemistry** -- Applying the principles of green chemistry in material selection and process design to reduce or eliminate the use and generation of hazardous substances.
- **Renewable Energy Integration** -- Designing products and systems to use renewable energy sources, and creating infrastructure to support renewable energy adoption.
- **Smart Systems and IoT** -- Leveraging smart technologies and the Internet of Things (IoT) to optimize resource use, enable predictive maintenance, and enhance overall system efficiency.
- **Sustainable Urban Design** -- Applying sustainability principles to urban planning and infrastructure design, creating cities that are more livable, resilient, and environmentally friendly.

18.8.4 Challenges in Sustainable Engineering Design

- **Balancing Trade-offs** -- Sustainable design often involves balancing competing priorities. For example, a more durable product might require more resources to produce initially but last longer, reducing overall resource consumption.
- **Measuring Sustainability** -- Quantifying the sustainability of a design can be challenging, particularly when considering long-term and indirect impacts.
- **Cost Considerations** -- Sustainable solutions may have higher upfront costs, even if they offer long-term savings. This can be a barrier to adoption, particularly in cost-sensitive markets.
- **Technological Limitations** -- Current technological limitations may constrain sustainable design options in some fields.

- **Regulatory Compliance** -- Navigating complex and sometimes conflicting regulations across different regions can be challenging for sustainable design implementation.
- **Changing Consumer Behavior** -- Many sustainable designs require changes in consumer behavior to be effective, which can be difficult to achieve.

18.8.5 Case Studies in Sustainable Engineering Design

- **Circular Smartphones** -- Companies like Fairphone are redesigning smartphones for modularity, repairability, and recyclability, extending product lifespans and reducing electronic waste.
- **Net-Zero Buildings** -- Advances in sustainable architecture are creating buildings that produce as much energy as they consume, using a combination of energy-efficient design, renewable energy systems, and smart technologies.
- **Sustainable Transportation** -- The development of electric vehicles, coupled with renewable energy charging infrastructure, represents a significant step towards sustainable transportation.
- **Biomimetic Wind** -- Turbines Wind turbine designs inspired by humpback whale fins have shown increased efficiency and reduced noise pollution.

18.8.6 The Future of Sustainability in Engineering Design

As we move forward, several trends are likely to shape the future of sustainable engineering design:

- **Artificial Intelligence and Machine Learning** -- AI and ML will play an increasingly important role in optimizing designs for sustainability, from material selection to predictive maintenance.
- **Advanced Materials** -- The development of new, sustainable materials (e.g., biodegradable plastics, carbon-negative concrete) will open up new possibilities for sustainable design.
- **Regenerative Design** -- Moving beyond "sustainable" to "regenerative" design that actively improves environmental and social conditions.
- **Systems Thinking** -- Increasingly complex global challenges will require engineers to adopt more holistic, systems-level approaches to sustainable design.
- **Interdisciplinary Collaboration** -- Sustainable engineering design will increasingly involve collaboration across disciplines, including environmental science, social science, economics, and policy.

Sustainability in engineering design is not just an ethical imperative; it's a practical necessity as we face unprecedented global challenges. By embracing sustainability principles, engineers can create designs that not only minimize negative impacts but actively contribute to environmental regeneration, economic prosperity, and social well-being.

The transition to truly sustainable engineering design requires a paradigm shift in how we approach problem-solving. It demands that we consider long-term impacts, embrace systems thinking, and prioritize regenerative solutions. While challenges remain, the potential for positive impact is immense.

As engineers, we have the opportunity and responsibility to be at the forefront of this transition. By integrating sustainability into every aspect of our designs, we can play a crucial role in creating a more sustainable, resilient, and equitable world. The future of our planet and society depends on our ability to innovate and implement sustainable solutions, making sustainability not just a consideration in engineering design, but its fundamental purpose.

18.10 Assignments

Assignment 18-1. Safety Considerations in Design – Analysis and Documentation.

Objective:

The objective of this assignment is for your design team to **analyze safety considerations** related to your product, ensure that it meets safety standards, and document how safety was incorporated into the design and build. You will also **discuss any safety aspects considered but not implemented** and provide recommendations for future improvements. This process ensures that your design aligns with both industry practices and user safety requirements.

Assignment Instructions:

Part 1: Identify Relevant Safety Considerations for Your Design

    Step 1: Review the Product's Intended Use
- **Describe how the product will be used** and identify possible user interactions.
- Consider both **typical and unintended uses** that could pose risks to the user, bystanders, or the environment.

    Deliverable:
- A **summary of product use cases** and potential safety risks based on misuse or incorrect operation.

    Step 2: Research Safety Standards and Regulations
- Identify relevant **industry standards, codes, or regulations** (e.g., OSHA, ISO, ANSI) that apply to your product.
- **List safety requirements** related to your product's operation, materials, or environment.

    Deliverable:
- A **list of safety standards and regulations** applicable to your design.

    Step 3: Develop a Safety Checklist
- Create a **checklist of safety considerations** for your product, organized into key categories. Consider:

- **Mechanical safety:** Avoiding pinch points, crush hazards, or structural failure
- **Electrical safety:** Grounding, insulation, and overload protection
- **Thermal safety:** Avoiding burns or thermal damage
- **Chemical safety:** Hazardous material handling or emissions
- **Environmental safety:** Noise, vibration, and pollution control
- **User safety:** Clear instructions, warnings, and fail-safe features

Deliverable:
- A **safety checklist** tailored to your product design.

Part 2: Design and Implement Safety Features

Step 1: Integrate Safety into the Design Process
- Identify the **design features that address specific safety considerations** from your checklist.

- **Examples:**
  - Guards or enclosures to prevent contact with moving parts
  - Overload protection to prevent electrical hazards
  - Non-toxic materials to ensure environmental safety
  - Labels and warnings to inform users of risks

Deliverable:
- A **table documenting safety features** and how each aligns with identified risks and requirements. Example:

Safety Risk	Design Feature Implemented	How it Addresses the Risk
Moving parts causing injury	Protective guard over gears	Prevents user contact with moving components
Overheating electronics	Thermal sensor and auto-shutoff	Stops operation when temperature exceeds safe limits

Step 2: Implement Safety in the Actual Build
- Ensure that the safety features were **properly incorporated during the build phase.**
- Document how safety mechanisms (e.g., guards, sensors) were physically implemented in the **final prototype or product build.**

Deliverable:
- **Photos or diagrams** showing where and how safety features were installed in the product.

Part 3: Evaluate Safety Features and Identify Gaps

Step 1: Test the Effectiveness of Safety Features
- Run tests to **evaluate the effectiveness** of the safety features.
  Example: Simulate situations where the safety features are supposed to engage (e.g., trigger an auto-shutoff or test guard placements).
- Record any **issues, failures, or unexpected behaviors** that arise during testing.

Step 2: Identify Safety Features Not Implemented
- Discuss any safety features that were considered but **not included in the final design** due to constraints (e.g., time, budget, or technical feasibility).
- Provide a **rationale for excluding these features** and assess their potential impact on product safety.

Deliverable:
- A **list of excluded safety features** with explanations for why they were not implemented.

Part 4: Recommend Future Improvements to the Design for Safety

Step 1: Reflect on Safety Testing Results
- **Analyze the results** from your safety tests. Were the implemented features sufficient?
- Identify any **additional safety risks** discovered during testing that were previously overlooked.

Step 2: Provide Recommendations for Future Safety Improvements
- Suggest **improvements or new features** that could enhance safety in future versions of the design.

- **Examples:**
  - Add redundant safety systems for critical operations
  - Improve visibility of warning labels
  - Use more durable materials to prevent part failure over time

Step 3: Reflect on the Safety Design Process
- If you were to restart the design process, **what would you do differently** regarding safety?
- Summarize any **lessons learned** about designing for safety, including the importance of early planning, user feedback, or testing methods.

Deliverable:
- A **Recommendations and Lessons Learned Report** (500-800 words) documenting:
  - Gaps identified during the design and testing process
  - Suggested improvements for future designs
  - Key takeaways from the entire safety design process

Part 5: Compile the Final Safety Report

Step 1: Organize Your Report
- Compile all your deliverables into a **Final Safety Report**, including:
  - Product use summary and potential risks
  - List of safety standards and regulations
  - Safety checklist and implemented design features
  - Photos or diagrams showing implemented safety features
  - List of excluded features and rationale
  - Test results and recommendations for improvements

Step 2: Submit Your Final Safety Report
- Ensure your report is **well-organized** and free of errors, with clear section headings.

Deliverable:
- A **PDF report** containing all documentation and images, submitted by the deadline.

Timeline and Suggested Schedule:
- **Week 1:** Identify safety considerations and research safety standards
- **Week 2:** Develop the safety checklist and document safety features in the design
- **Week 3:** Implement safety features in the product build and run safety tests
- **Week 4:** Finalize recommendations and compile the final report

Evaluation Criteria:
- **Completeness:** Were all parts of the assignment completed?
- **Quality of Safety Analysis:** Were relevant safety risks identified and addressed?
- **Effectiveness of Safety Features:** Were implemented features appropriate and functional?
- **Documentation:** Was the safety design process well-documented?
- **Recommendations:** Were meaningful recommendations provided for future improvements?
- **Professionalism:** Is the final report clear, well-organized, and professionally presented?

Assignment 18-2. Ergonomics Considerations in Design – Analysis and Documentation.

Objective:

This assignment requires your design team to **analyze and incorporate ergonomics** in your product. You will evaluate various ergonomics principles, document how they were integrated into the design, and discuss any ergonomic considerations that were omitted. Finally, you will **suggest recommendations for future improvements** and reflect on lessons learned during the process.

Assignment Instructions:

Part 1: Identify Ergonomics Aspects Relevant to Your Design

    Step 1: Analyze User Interaction with the Product
- Identify how users will **interact with the product** (e.g., lifting, handling, operating controls, sitting).
- Consider all **potential users** (different body sizes, strengths, and abilities). Ensure your design **accommodates a diverse group of users**, including left- and right-handed individuals, people of varying heights, and those with disabilities.

    Deliverable:
- A **summary of user interaction scenarios** and user profiles.

    Step 2: Research Ergonomics Standards and Guidelines
- Identify **ergonomic guidelines** and standards (e.g., ANSI/HFES, ISO ergonomics standards) relevant to your product.
- Review **best practices** such as:
  - Anthropometric data (e.g., hand sizes, reach distances, height ranges)
  - Optimal force levels for operation
  - Comfortable working postures

    Deliverable:
- A **list of relevant ergonomics standards** and guidelines applicable to the

product.

Part 2: Design for Ergonomics

    Step 1: Identify Key Ergonomics Considerations
- List the **key ergonomics considerations** applicable to your product.
  Examples:
  - Proper **posture** while using the product
  - **Ease of use** of controls or interfaces
  - **Weight distribution** and handling for portable products
  - **Accessibility** for users with disabilities

    Deliverable:
- A **list of ergonomic principles** applied in your design.

    Step 2: Implement Ergonomic Features in the Design
- Document the **ergonomic features incorporated** into your design.
  Examples:
  - Adjustable seats, levers, or handles
  - Proper spacing and size of buttons to avoid user fatigue
  - Rounded edges to prevent injury
- Explain **how these features enhance user comfort and safety**.

    Deliverable:
- A **table of ergonomic features** with their purpose. Example:

Ergonomic Principle	Feature Implemented	How it Improves Usability
Reduce user fatigue	Padded grip handle	Improves comfort and reduces strain on hand muscles
Accessible controls	Large tactile buttons	Easier to use for people with limited dexterity

    Step 3: Implement Ergonomics in the Build Process
- Ensure that ergonomic features were properly integrated during the **actual product build**.
- Provide **visual evidence** (e.g., photos, diagrams) of ergonomic features in the built product.

    Deliverable:
- **Photos or diagrams** showing ergonomic elements in the final product.

Part 3: Evaluate Ergonomics and Identify Gaps

Step 1: Test Ergonomic Features with Users or Simulations
- **Test the ergonomic features** by simulating use cases or involving real users in usability tests.
- Record any **user feedback or challenges** encountered during testing.
- Analyze whether users experienced **comfort, fatigue, or difficulty** interacting with the product.

Deliverable:
- A **summary of test results** and user feedback.

Step 2: Identify Ergonomic Features Not Implemented
- List any **ergonomic improvements** that were considered but **not implemented** due to constraints (e.g., time, budget, manufacturing limitations).
- Discuss the **impact** of excluding these features on user experience.

Deliverable:
- A **list of omitted ergonomic features** with explanations for their exclusion.

Part 4: Recommendations for Future Ergonomics Improvements

Step 1: Reflect on Ergonomics Testing and User Feedback
- Assess whether the **ergonomic features met expectations** during testing.
- Identify any **additional ergonomic improvements** needed for future iterations of the design.

Step 2: Provide Recommendations for Future Design Enhancements
- Suggest **practical improvements** for future designs.
  Examples:
  - Adding more **adjustability** to seats or handles
  - Reducing **weight** to enhance portability
  - Improving **user interface design** for better accessibility

Step 3: Reflect on Lessons Learned
- If you could restart the design process, **what ergonomic changes would**

18.10 Assignments

you make?
- o Summarize **key lessons learned** about designing for ergonomics, including the importance of early user involvement and testing.

Deliverable:
- o A **Recommendations and Lessons Learned Report** (500-800 words) detailing:
  - Feedback from ergonomics testing
  - Suggested improvements for future iterations
  - Key takeaways from the design process

Part 5: Compile the Final Ergonomics Report

Step 1: Organize Your Report
- o Compile all your documentation into a **Final Ergonomics Report**, including:
  - Summary of user interaction and ergonomics guidelines
  - List of ergonomic features implemented
  - Photos or diagrams showing ergonomic elements in the build
  - List of omitted ergonomic features with explanations
  - Recommendations and lessons learned

Step 2: Submit Your Final Ergonomics Report
- o Ensure the report is **well-organized and error-free**, with clear section headings.

Deliverable:
- o A **PDF report** containing all required elements, submitted by the deadline.

Timeline and Suggested Schedule:
- **Week 1:** Analyze user interaction and develop ergonomic principles
- **Week 2:** Implement ergonomic features and test the design
- **Week 3:** Collect user feedback and assess ergonomics performance
- **Week 4:** Document findings, recommend improvements, and submit the final report

Evaluation Criteria:
- **Completeness:** Were all steps of the assignment completed and documented?
- **Quality of Ergonomics Design:** Were appropriate ergonomic principles applied in the design?
- **Testing and Feedback:** Was testing thorough and documented with meaningful feedback?
- **Recommendations:** Were thoughtful recommendations provided for future improvements?
- **Professionalism:** Is the final report clear, organized, and polished?

Assignment 18-3. Health Impact Considerations in Design – Analysis and Documentation.

Objective:
This assignment requires your design team to **analyze health impacts** associated with your product and document how health-related features were incorporated into the design and build. You will assess potential risks to users' physical and mental health, outline how your design mitigates these risks, and discuss health-related improvements that were considered but not implemented. Finally, you will recommend future improvements based on your findings.

Assignment Instructions:

Part 1: Identify Relevant Health Impact Aspects

Step 1: Analyze Product Use and Potential Health Risks
- Review how the product will be **used by users** and identify any **health risks** associated with its use.

  Consider:
  - **Physical health impacts:** Risks such as repetitive strain, exposure to harmful substances, noise, or heat.
  - **Mental health impacts:** Cognitive load, stress from complex interfaces, or fatigue from prolonged use.

Deliverable:
- A **summary of potential health impacts** associated with the product (both physical and mental health aspects).

Step 2: Research Health-Related Standards and Guidelines
- Identify any **health-related standards** (e.g., OSHA, WHO, ISO) relevant to your product.
- List best practices such as:
  - Reducing exposure to hazardous materials
  - Minimizing noise levels to avoid hearing damage
  - Designing to prevent repetitive strain injuries

Deliverable:
- A **list of applicable health-related standards and best practices** that guide your product's design.

Part 2: Design Health Features and Implement in the Build

Step 1: Identify Health-Oriented Design Features
- Document **health features** incorporated into your design to reduce risks and enhance user well-being.

- **Examples:**
  - Ergonomic handles to reduce repetitive strain injuries
  - Noise-dampening materials to minimize sound exposure
  - Built-in timers to reduce user fatigue by encouraging breaks

Deliverable:
- A **table of health features** and their purpose. Example:

Health Risk	Design Feature Implemented	How it Reduces Risk
Repetitive strain on hands	Soft-grip ergonomic handles	Reduces pressure on muscles and tendons
Prolonged exposure to noise	Noise-dampening panels	Prevents hearing damage from high noise
Cognitive fatigue from long use	Usage timer with reminders	Encourages breaks to improve focus

Step 2: Incorporate Health Features into the Build
- Ensure that the health-related features were **properly integrated into the final product build**.
- Document the **installation or fabrication of these features** using sketches, photos, or diagrams.

Deliverable:
- **Photos or diagrams** showing health features incorporated in the build.

Part 3: Evaluate Health Features and Identify Gaps

Step 1: Test the Effectiveness of Health Features

18.10 Assignments

- Conduct usability tests to **evaluate the effectiveness of health features**.
- Simulate user interactions to **verify whether the product supports user health** as intended.

- **Examples:**
  - Use a grip strength test to ensure ergonomic handles reduce muscle strain.
  - Measure noise levels with a sound meter to confirm compliance with hearing protection guidelines.

Deliverable:
- A **summary of test results** showing how the health features performed in real use cases.

Step 2: Identify Health Features Not Implemented
- List **health features that were considered but not included** in the final design due to constraints (e.g., budget, time, or technical feasibility).
- Discuss the **impact of excluding these features** on user health.

Deliverable:
- A **list of omitted health features** with reasons for their exclusion.

Part 4: Recommend Future Health-Oriented Improvements

Step 1: Reflect on Health Feature Testing and User Feedback
- Analyze the **feedback and test results**. Were the health features effective in promoting well-being?
- Identify any **additional health-related improvements** needed based on the test outcomes.

Step 2: Provide Recommendations for Future Health Improvements
- Suggest **new health-oriented features or enhancements** for future product iterations.

- **Examples:**
  - Incorporate real-time health monitoring for users (e.g., heart rate sensors)
  - Use lighter materials to reduce strain during product handling
  - Improve visual interfaces to reduce eye strain

Step 3: Reflect on the Health Impact Design Process
- If you were to start the design process again, **what would you do differently** regarding health considerations?
- Summarize the **lessons learned** from designing for health, including the importance of user feedback and early testing.

Deliverable:
- A **Recommendations and Lessons Learned Report** (500-800 words) summarizing:
  - Test results and feedback on health features
  - Suggested improvements for future designs
  - Key takeaways from the entire design process

Part 5: Compile the Final Health Impact Report

Step 1: Organize Your Report
- Compile all documentation into a **Final Health Impact Report**, including:
  - Summary of health risks and standards
  - List of health features implemented
  - Photos or diagrams showing health features in the build
  - List of omitted features with explanations
  - Recommendations and lessons learned

Step 2: Submit Your Final Report
- Ensure your report is **well-organized and professionally formatted** with clear section headings.

Deliverable:
- A **PDF report** containing all required elements, submitted by the deadline.

Timeline and Suggested Schedule:
- **Week 1:** Analyze health risks and research relevant standards
- **Week 2:** Implement health features and conduct usability tests
- **Week 3:** Collect user feedback and assess the effectiveness of health features
- **Week 4:** Finalize recommendations and compile the final report

Evaluation Criteria:
- **Completeness:** Were all required sections of the assignment completed?
- **Quality of Health Feature Design:** Were appropriate health risks identified and mitigated?
- **Effectiveness of Testing:** Were usability tests well-designed, and were results meaningful?
- **Recommendations:** Were thoughtful recommendations provided for future health improvements?
- **Professionalism:** Is the final report clear, organized, and well-presented?

Assignment 18-4. Environmental Impact Considerations in Design – Analysis and Documentation.

Objective:
This assignment requires your design team to **analyze environmental impacts** associated with your product and document how environmental protection features were incorporated into the design and build. You will assess potential environmental risks, outline how your design mitigates these impacts, and discuss environmental features that were considered but not implemented. Finally, you will recommend future improvements based on your findings.

Assignment Instructions:

Part 1: Identify Environmental Impact Aspects Relevant to Your Design

    Step 1: Analyze the Product Life Cycle and Potential Environmental Impacts
- Review the entire **life cycle of the product** (from raw materials to disposal) and identify potential **environmental impacts** at each stage. **Consider the following life cycle stages:**
    - **Material sourcing:** Are the materials renewable, recyclable, or sustainably sourced?
    - **Manufacturing process:** Does the production method use excess energy, water, or produce harmful emissions?
    - **Use phase:** Does the product consume high amounts of energy or release harmful by-products during use?
    - **End-of-life disposal:** Is the product recyclable, biodegradable, or will it contribute to waste?

    Deliverable:
- A **summary of environmental impacts** associated with each stage of the product life cycle, highlighting potential risks such as resource depletion, pollution, and energy consumption.

    Step 2: Research Environmental Standards and Regulations
- Identify relevant **environmental standards and regulations** that apply to your product (e.g., ISO 14001 for environmental management, RoHS for hazardous materials, Energy Star for energy efficiency).

- o Outline any **guidelines** for sustainability, resource efficiency, and waste reduction that inform your product design.

Deliverable:
- o A **list of environmental standards, regulations, and best practices** that apply to the design of your product.

Part 2: Design and Implement Environmental Protection Features

Step 1: Identify Key Environmental Protection Features
- o Document **environmentally-friendly features** that were incorporated into your design to mitigate the identified risks.

- o **Examples:**
    - Using **recycled or biodegradable materials**
    - Reducing energy consumption through **efficient components**
    - Minimizing waste by designing for **easy disassembly and recycling**
    - **Modular design** to facilitate repair and extend the product's lifespan

Deliverable:
- o A **table of environmental protection features** with an explanation of how each one reduces environmental impact. Example:

Environmental Impact	Design Feature Implemented	How it Reduces Impact
Non-recyclable materials	Use of recycled aluminum	Reduces reliance on raw material extraction
High energy consumption	Energy-efficient motor	Lowers overall energy use during product use
End-of-life disposal waste	Modular design for recycling	Allows for easy disassembly and material recovery

Step 2: Implement Environmental Features in the Build
- o Ensure that these environmental features were properly integrated during the **actual product build**.
- o Document the **implementation of these features** using photos, diagrams, or sketches.

Deliverable:
- **Photos or diagrams** showing the incorporation of environmental features in the final product build.

Part 3: Evaluate Environmental Features and Identify Gaps

Step 1: Test and Assess Environmental Features
- Evaluate the effectiveness of the environmental protection features through **testing or simulations**.
- **Measure performance** in areas such as energy efficiency, material sustainability, or recyclability.

- **Examples:**
  - Use energy meters to assess power consumption.
  - Conduct material tests to verify the recyclability or biodegradability of components.

Deliverable:
- A **summary of test results** showing the performance of environmental protection features.

Step 2: Identify Environmental Features Not Implemented
- List any **environmental protection features that were considered but not included** in the final design due to constraints (e.g., cost, manufacturing limitations, technical feasibility).
- Discuss the **impact of not implementing these features** on the product's environmental footprint.

Deliverable:
- A **list of omitted environmental features** with reasons for their exclusion.

Part 4: Recommend Future Environmental Improvements

Step 1: Reflect on Environmental Testing and Results
- Analyze the **results from testing**. Did the environmental protection features perform as expected? Were there any issues or room for improvement?

18.10 Assignments

- Identify any **additional environmental considerations** that need to be addressed.

Step 2: Provide Recommendations for Future Design Enhancements
- Suggest **practical improvements** or new environmental features for future iterations of the product.

- **Examples:**
  - Use of more sustainable materials or reducing the carbon footprint of manufacturing
  - Improving energy efficiency by incorporating advanced technology
  - Designing for **product take-back programs** or offering repair services to reduce waste

Step 3: Reflect on the Environmental Design Process
- If you were to restart the design process, **what would you do differently** in terms of environmental considerations?
- Summarize the **lessons learned** from designing for environmental protection, including the importance of early consideration of environmental impacts and continuous improvement.

Deliverable:
- A **Recommendations and Lessons Learned Report** (500-800 words) that:
  - Summarizes the performance of environmental features
  - Provides suggestions for future improvements
  - Reflects on key takeaways from the design process

Part 5: Compile the Final Environmental Impact Report

Step 1: Organize Your Report
- Compile all documentation into a **Final Environmental Impact Report**, including:
  - Summary of the product's life cycle and environmental impacts
  - List of environmental standards and regulations considered
  - Documentation of environmental protection features implemented
  - Photos or diagrams showing features in the build
  - List of omitted environmental features and rationale

- Recommendations for future improvements

Step 2: Submit Your Final Environmental Impact Report
- Ensure the report is **well-organized, professionally formatted**, and includes clear section headings for each part of the assignment.

Deliverable:
- A **PDF report** containing all required elements, submitted by the deadline.

Timeline and Suggested Schedule:
- **Week 1:** Analyze product life cycle and research environmental standards
- **Week 2:** Design and implement environmental protection features
- **Week 3:** Test and evaluate the performance of environmental features
- **Week 4:** Document findings, recommend improvements, and submit the final report

---

Evaluation Criteria:
- **Completeness:** Were all steps of the assignment completed and documented?
- **Quality of Environmental Design:** Were appropriate environmental protection features incorporated and tested?
- **Effectiveness of Testing:** Were the environmental features thoroughly tested and evaluated?
- **Recommendations:** Were thoughtful recommendations provided for future environmental improvements?
- **Professionalism:** Is the final report well-organized, clearly written, and professionally presented?

18.10 Assignments

Assignment 18-5. Social Impact Considerations in Design – Analysis and Documentation.

Objective:
The goal of this assignment is for your design team to **analyze the social impacts** associated with your product and document how social impact features were integrated into the design and build. You will assess potential positive and negative social impacts, outline design choices that address these considerations, and discuss aspects that were considered but not implemented. Finally, you will recommend future improvements to enhance the product's social impact.

Assignment Instructions:

Part 1: Identify Social Impact Aspects Relevant to Your Design

   Step 1: Analyze the Product's Social Impact
   - Identify the **intended and unintended social impacts** of your product on different stakeholders (users, communities, and society at large).

   Consider:
   - **Accessibility:** Does the product cater to users with varying abilities?
   - **Affordability:** Can a broad range of people afford the product?
   - **User diversity:** Does the product meet the needs of users across different demographics (age, gender, ethnicity)?
   - **Job creation or displacement:** Does the product promote employment, or could it lead to job loss through automation?
   - **Social well-being:** How does the product improve users' quality of life?
   - **Community impact:** Does the product support or disrupt local communities?

   Deliverable:
   - A **summary of potential social impacts** (both positive and negative) associated with the product.

   Step 2: Research Social Impact Frameworks and Guidelines
   - Identify relevant **social impact frameworks or best practices** (e.g., UN

Sustainable Development Goals, ISO 26000 for social responsibility).
- o **Consider social equity and justice** principles, such as fair labor practices, inclusivity, and ethical production.

Deliverable:
- o A **list of relevant social frameworks and guidelines** that inform your product design.

Part 2: Design and Implement Social Impact Features

Step 1: Identify Key Social Impact Features
- o Document the **socially responsible features** incorporated into your product to address the identified social impact aspects.

- o **Examples:**
  - **Adjustable or modular components** to accommodate diverse user needs.
  - **Use of affordable materials** to ensure product accessibility across different economic levels.
  - **Design for inclusivity**, such as language-free interfaces or features for individuals with disabilities.
  - **Ethical sourcing** of materials to promote fair labor practices.

Deliverable:
- o A **table of social impact features** with their purpose. Example:

Social Impact Consideration	Feature Implemented	How it Addresses the Impact
Accessibility for all users	Modular seat adjustment	Accommodates users of different sizes and abilities
Affordability	Use of recycled materials	Lowers production cost, making product affordable
Support for marginalized groups	Braille on controls	Improves usability for visually impaired individuals

Step 2: Implement Social Impact Features in the Build
- o Integrate these features into your **prototype or product build**.
- o Provide **visual documentation** (photos, sketches, or CAD models) showing how the social impact features were incorporated into the final product.

Deliverable:
- **Photos or diagrams** illustrating the social impact features in the built product.

Part 3: Evaluate Social Impact Features and Identify Gaps

Step 1: Assess the Effectiveness of Social Impact Features
- Conduct **usability tests or feedback sessions** with a diverse group of users to evaluate the social impact features.
- Document any **user feedback or challenges** experienced during testing, particularly from marginalized or underrepresented groups.

Deliverable:
- A **summary of test results and user feedback**, focusing on how effectively the product meets social impact goals.

Step 2: Identify Social Impact Features Not Implemented
- List **social impact features that were considered but not included** in the final design due to constraints (e.g., budget, time, or technical feasibility).
- Explain the **potential impact** of not including these features.

Deliverable:
- A **list of omitted social features** with an explanation for their exclusion.

Part 4: Recommend Future Social Impact Improvements

Step 1: Reflect on the Testing Results and Feedback
- Analyze the **feedback from users** and identify any **gaps or shortcomings** in the social impact of the product.
- Determine whether the current design promotes **inclusivity, affordability, and well-being** as intended.

Step 2: Provide Recommendations for Future Design Improvements
- Suggest **improvements or new features** to enhance the product's social impact in future iterations.
  Examples:
  - Expand the range of user adjustability features for greater inclusivity.

- Improve accessibility by adding tactile feedback or audio interfaces.
- Partner with community organizations to **promote product use among underserved populations**.

Step 3: Reflect on the Social Impact Design Process
- If you were to restart the design process, **what would you change** to improve the social impact?
- Summarize **lessons learned** about designing for social impact, such as the importance of engaging with users early in the design process or aligning with ethical production practices.

Deliverable:
- A **Recommendations and Lessons Learned Report** (500-800 words) summarizing:
  - Test results and feedback on social impact features
  - Suggestions for future improvements
  - Key takeaways from the design process

Part 5: Compile the Final Social Impact Report

Step 1: Organize Your Report
- Compile all documentation into a **Final Social Impact Report**, including:
  - Summary of social impacts and relevant frameworks
  - Documentation of social impact features and their purpose
  - Photos or diagrams showing implemented features
  - List of omitted features with explanations
  - Recommendations and lessons learned

Step 2: Submit Your Final Social Impact Report
- Ensure your report is **well-organized and free of errors**, with clear section headings for each part of the assignment.

Deliverable:
- A **PDF report** containing all required elements, submitted by the deadline.

Timeline and Suggested Schedule:

- **Week 1:** Identify social impacts and research relevant frameworks
- **Week 2:** Design and implement social impact features
- **Week 3:** Conduct usability tests and gather feedback
- **Week 4:** Document findings, recommend improvements, and submit the final report

Evaluation Criteria:
- **Completeness:** Were all required sections completed and documented?
- **Quality of Social Impact Design:** Were meaningful social impact features incorporated and tested?
- **Effectiveness of Testing and Feedback:** Were the tests thorough, and was user feedback addressed?
- **Recommendations:** Were thoughtful recommendations provided for future improvements?
- **Professionalism:** Is the final report clear, organized, and well-presented?

Assignment 18-6. Ethical Considerations in Design – Analysis and Documentation.

Objective:

This assignment requires your design team to **analyze and address ethical considerations** associated with your product. You will document how ethical principles were incorporated into the design and build, discuss aspects that were considered but not implemented, and provide recommendations for future improvements. Ethical design ensures the product aligns with professional standards, promotes fairness, and avoids harm to users, society, and the environment.

Assignment Instructions:

Part 1: Identify Ethical Considerations in the Product Design

Step 1: Analyze Ethical Risks and Responsibilities
- Identify **ethical risks** and responsibilities related to the product's use and design. Consider:
  - **Safety and harm prevention:** Does the product minimize risks to users and bystanders?
  - **Privacy and data security:** Does the product collect personal data, and how is it protected?
  - **Fairness and inclusivity:** Does the product serve all intended users fairly, without bias or exclusion?
  - **Environmental ethics:** Does the product promote sustainable practices?
  - **Transparency:** Are the product's limitations, risks, and usage instructions clearly communicated?
  - **Social responsibility:** Does the product contribute positively to society?

Deliverable:
- A **summary of ethical considerations** and potential risks related to the design and use of the product.

Step 2: Research Ethical Guidelines and Standards
- Review **ethical frameworks and professional codes of ethics** that apply to your field (e.g., IEEE Code of Ethics, NSPE Code of Ethics for

18.10 Assignments

Engineers, ISO 26000 for social responsibility).
- Identify **best practices for ethical product design** based on these standards.

Deliverable:
- A **list of ethical frameworks and standards** that inform your design decisions.

Part 2: Design for Ethical Considerations

Step 1: Identify Ethical Design Features
- Document **design features that address the identified ethical risks**. **Examples:**
  - Safety features to prevent harm (e.g., emergency shutoff switches).
  - Data encryption to protect user privacy.
  - Accessible design to ensure fair access to people with disabilities.
  - Clear user instructions and warnings to promote informed use.

Deliverable:
- A **table of ethical design features** with an explanation of how each one mitigates risks. Example:

Ethical Concern	Design Feature Implemented	How it Addresses the Concern
User safety	Emergency shutoff button	Prevents accidents by allowing users to quickly stop the product in emergencies
Privacy and data protection	Encrypted data storage	Secures user information and prevents unauthorized access
Accessibility	Voice command interface	Ensures the product can be used by individuals with limited mobility or vision

Step 2: Implement Ethical Features in the Build
- Ensure that the ethical features were **integrated into the prototype or product build**.
- Document how these features were **physically implemented** or configured in the product.

Deliverable:
- **Photos, sketches, or diagrams** illustrating how ethical design features were integrated into the product.

Part 3: Evaluate Ethical Features and Identify Gaps

    Step 1: Assess the Effectiveness of Ethical Features
- Test or **simulate product usage** to evaluate the performance of the ethical features.
- Gather **feedback from users or stakeholders** to determine whether the product meets ethical expectations and prevents harm.

    Deliverable:
- A **summary of test results and user feedback** focusing on the ethical performance of the product.

    Step 2: Identify Ethical Aspects Not Implemented
- List any **ethical considerations that were explored but not included** in the final design due to constraints (e.g., budget, time, or technical challenges).
- Discuss the **potential risks** or impacts of not implementing these features.

    Deliverable:
- A **list of omitted ethical aspects** with an explanation for their exclusion.

Part 4: Recommend Future Ethical Improvements

    Step 1: Reflect on the Ethical Testing and Feedback
- Analyze the **results from testing**. Were the ethical features effective? Did any unexpected issues arise?
- Identify any **additional ethical concerns** that need to be addressed in future product versions.

    Step 2: Provide Recommendations for Future Design Improvements
- Suggest **practical improvements** for future versions of the product.
  Examples:
  - Enhance user privacy by minimizing the collection of personal data.
  - Improve transparency by providing more detailed user instructions and warnings.
  - Incorporate sustainability features to reduce environmental impact.

Step 3: Reflect on the Ethical Design Process
- If you were to **restart the design process**, what ethical aspects would you approach differently?
- Summarize the **lessons learned** about ethical design and discuss how these lessons will guide future projects.

Deliverable:
- A **Recommendations and Lessons Learned Report** (500-800 words) that:
  - Summarizes feedback and testing outcomes.
  - Suggests future improvements.
  - Reflects on key takeaways from the ethical design process.

Part 5: Compile the Final Ethical Impact Report

Step 1: Organize Your Report
- Compile all deliverables into a **Final Ethical Impact Report**, including:
  - Summary of ethical risks and responsibilities.
  - List of ethical frameworks and standards used.
  - Documentation of ethical features implemented.
  - Photos or diagrams of the product build.
  - List of omitted ethical aspects with explanations.
  - Recommendations and lessons learned.

Step 2: Submit Your Final Ethical Impact Report
- Ensure the report is **well-organized and polished**, with clear section headings and a professional format.

Deliverable:
- A **PDF report** containing all required elements, submitted by the deadline.

Timeline and Suggested Schedule:
- **Week 1:** Identify ethical risks and research relevant frameworks.
- **Week 2:** Design and implement ethical features in the product.
- **Week 3:** Test and evaluate the performance of ethical features.
- **Week 4:** Document findings, recommend improvements, and submit the final report.

Evaluation Criteria:
- **Completeness:** Were all parts of the assignment completed and documented?
- **Quality of Ethical Design:** Were ethical risks identified, and were appropriate features incorporated?
- **Effectiveness of Testing:** Were the ethical features tested thoroughly, and was feedback meaningful?
- **Recommendations:** Were thoughtful recommendations provided for future ethical improvements?
- **Professionalism:** Is the final report clear, well-organized, and professionally presented?

Assignment 18-7. Political Considerations in Design – Analysis and Documentation.

Objective:
This assignment requires your design team to **analyze political considerations** that may affect your product and document how these were incorporated into the design and build. Political factors can influence how a product is developed, regulated, distributed, and perceived in different regions or markets. You will also identify political aspects that were considered but not implemented, and recommend improvements for future iterations of the product.

Assignment Instructions:

Part 1: Identify Political Considerations Relevant to Your Design

    Step 1: Analyze the Political Environment
- Research the **political landscape** in which the product will be developed, sold, or distributed.

    Consider the following:
- **Regulatory frameworks** (local, national, or international laws)
- **Compliance with trade policies** (e.g., import/export restrictions, tariffs)
- **Government incentives or subsidies** for sustainable products
- **Political stability** and its influence on supply chains or market demand
- **Intellectual property laws** and licensing regulations
- **Geopolitical considerations**, such as international trade conflicts or sanctions

    Deliverable:
- A **summary of political factors** relevant to your product and its deployment.

    Step 2: Identify Relevant Political Guidelines and Policies
- Investigate any **laws, policies, or political initiatives** that your design must align with.

- **Examples:**

- GDPR (General Data Protection Regulation) for privacy compliance in the EU
- RoHS (Restriction of Hazardous Substances) directive for electronic products
- Government policies promoting renewable energy adoption

Deliverable:
- A **list of political policies and regulations** that your product needs to comply with, along with a brief description of each.

Part 2: Design and Implement Political Considerations in the Product

Step 1: Identify Political Design Features
- Document **design features or decisions** made to align with political requirements.

- **Examples:**
    - Selecting **materials that meet trade restrictions** (e.g., avoiding conflict minerals).
    - **Complying with environmental regulations** by incorporating low-emission technologies.
    - Ensuring **data privacy compliance** through encrypted communication protocols.
    - Designing the product to meet **energy-efficiency standards** mandated by governments.

Deliverable:
- A **table of political design features** and their purpose. Example:

Political Consideration	Design Feature Implemented	How it Aligns with Political Requirements
GDPR compliance	Encrypted user data storage	Ensures user privacy in compliance with the GDPR
Trade restrictions on materials	Use of certified suppliers	Avoids using restricted raw materials like conflict minerals
Energy efficiency mandates	Integrated low-power components	Reduces energy consumption to meet government standards

Step 2: Implement Political Considerations in the Product Build
- Ensure the identified political features were integrated into the **prototype or final build**.

18.10 Assignments

- Document how the design team ensured **compliance with political requirements** during the manufacturing or assembly stages.

Deliverable:
- **Photos, diagrams, or documentation** showing how political features were incorporated into the final product.

Part 3: Evaluate Political Features and Identify Gaps

Step 1: Assess the Effectiveness of Political Features
- Evaluate whether the political features implemented in your product **align with required policies or regulations**.
- Test whether the product is **ready for certification or approval** by relevant government bodies (e.g., CE marking, Energy Star certification).

Deliverable:
- A **summary of results** from compliance testing or certification processes.

Step 2: Identify Political Aspects Not Implemented
- List any **political considerations that were explored but not implemented** due to constraints (e.g., cost, feasibility, or limited resources).
- Discuss the **impact of not including these features** on the product's political or market viability.

Deliverable:
- A **list of omitted political considerations** with an explanation for their exclusion.

Part 4: Recommend Future Political Improvements

Step 1: Reflect on the Testing Results and Feedback
- Analyze whether the **product aligns with all political expectations**. Were any risks or compliance gaps discovered during testing?
- Identify any **additional political considerations** that should be addressed in future product versions.

Step 2: Provide Recommendations for Future Political Improvements

- Suggest **design or process improvements** to better align the product with political requirements in future iterations.

- **Examples:**
  - Adjust the product to **meet future regulatory changes** (e.g., stricter environmental policies).
  - Engage with **government programs or incentives** to promote product adoption.
  - Improve the **sourcing strategy** to minimize risks from geopolitical events.

Step 3: Reflect on the Political Design Process
- If you were to **restart the design process**, what would you do differently to address political factors?
- Summarize the **lessons learned** from incorporating political considerations, such as the importance of early policy research or stakeholder engagement.

Deliverable:
- A **Recommendations and Lessons Learned Report** (500-800 words) that:
  - Summarizes the compliance outcomes and feedback.
  - Provides actionable recommendations for future political improvements.
  - Reflects on the key takeaways from the political design process.

Part 5: Compile the Final Political Impact Report

Step 1: Organize Your Report
- Compile all documentation into a **Final Political Impact Report**, including:
  - Summary of political factors and relevant policies.
  - Documentation of political features implemented.
  - Photos or diagrams showing political features in the build.
  - List of omitted political considerations with explanations.
  - Recommendations and lessons learned.

Step 2: Submit Your Final Political Impact Report
- Ensure your report is **well-organized and polished**, with clear section

headings and professional formatting.

Deliverable:
- A **PDF report** containing all required elements, submitted by the deadline.

Timeline and Suggested Schedule:
- **Week 1:** Identify political considerations and research relevant policies.
- **Week 2:** Design and implement political features in the product.
- **Week 3:** Test compliance and gather feedback.
- **Week 4:** Document findings, recommend improvements, and submit the final report.

Evaluation Criteria:
- **Completeness:** Were all steps of the assignment completed and documented?
- **Quality of Political Design:** Were appropriate political factors identified, and were meaningful features implemented?
- **Effectiveness of Testing:** Was the product evaluated for compliance, and were meaningful feedback or risks identified?
- **Recommendations:** Were thoughtful recommendations provided for future political improvements?
- **Professionalism:** Is the final report well-organized, clear, and professionally presented?

Assignment 18-8. Sustainability Considerations in Design – Analysis and Documentation.

Objective:
The objective of this assignment is for your design team to **analyze sustainability considerations** associated with your product and document how these were incorporated into the design and build. You will evaluate the environmental, economic, and social sustainability of the product and identify design decisions that address these aspects. Additionally, you will discuss any sustainability features that were considered but not implemented, and recommend improvements for future versions.

Assignment Instructions:

Part 1: Identify Sustainability Aspects Relevant to Your Design

    Step 1: Define Sustainability in the Context of Your Product
- Begin by researching **sustainability frameworks** (e.g., **United Nations Sustainable Development Goals (SDGs), Circular Economy Principles**) relevant to your product.
- Consider the **triple bottom line** of sustainability:
  - **Environmental sustainability:** Minimizing waste, emissions, and resource consumption.
  - **Economic sustainability:** Reducing costs over the product life cycle, promoting long-term usability, and supporting local economies.
  - **Social sustainability:** Promoting equity, fair labor practices, and positive community impacts.

    Deliverable:
- A **summary of sustainability considerations** for your product under the three categories: environmental, economic, and social.

    Step 2: Identify Key Sustainability Metrics
- Define **measurable metrics** to assess the sustainability of your product, such as:
  - **Carbon footprint** (CO_2 emissions from materials, production, and shipping)

- **Energy consumption** (during manufacturing and product use)
- **Recyclability** and **reuse** potential of materials
- **Longevity and durability** (extending product lifespan)
- **Water footprint** (amount of water used in production)

Deliverable:
- A **list of key sustainability metrics** with a description of how they apply to your design.

Part 2: Design for Sustainability and Implement in the Build

Step 1: Integrate Sustainability Features into the Design
- Identify **specific design features** that promote sustainability. Examples include:
    - **Material selection:** Use of **recycled or biodegradable materials**.
    - **Energy-efficient design:** Low-power components or energy-saving features during use.
    - **Modularity and repairability:** Design for easy **disassembly** and repair to extend the product's lifespan.
    - **Packaging:** Use of **minimal or eco-friendly packaging** to reduce waste.
    - **Local sourcing:** Sourcing materials or components from **local suppliers** to reduce transportation emissions.

Deliverable:
- A **table of sustainability features** with their impact on the product. Example:

Sustainability Aspect	Design Feature Implemented	How it Promotes Sustainability
Material reuse	Use of recycled aluminum	Reduces raw material extraction and energy use
Energy efficiency	Low-power LED lighting	Reduces energy consumption during product use
Modularity for repairability	Replaceable battery module	Extends product lifespan and reduces e-waste

Step 2: Implement Sustainability Features in the Build
- Integrate these sustainability features into the **actual build** of the product.
- Document the **steps taken** to implement these features during

manufacturing, assembly, or sourcing.

Deliverable:
- **Photos, diagrams, or CAD models** showing how sustainability features were implemented.

Part 3: Evaluate Sustainability Performance and Identify Gaps

Step 1: Assess the Effectiveness of Sustainability Features
- **Test or evaluate the performance** of sustainability features.
  Examples:
    - Measure **energy consumption** during product use to validate energy efficiency claims.
    - Conduct a **recyclability analysis** to determine how much of the product can be reused or recycled.
    - Perform a **life cycle assessment (LCA)** to calculate the environmental impact of the product from cradle to grave.

Deliverable:
- A **summary of sustainability performance tests** with key findings (e.g., energy savings achieved, materials successfully recycled).

Step 2: Identify Sustainability Features Not Implemented
- List **sustainability aspects or features that were considered but not included** in the final product.
- Provide the **rationale for excluding these features** (e.g., cost, technical feasibility, time constraints).
- Discuss the **potential impact of these omissions** on the product's overall sustainability.

Deliverable:
- A **list of omitted sustainability features** with an explanation for their exclusion.

Part 4: Recommend Future Sustainability Improvements

Step 1: Reflect on Sustainability Performance and Feedback
- Analyze the **test results and feedback** from stakeholders.
- Identify **gaps** in the current design and areas where sustainability features could be enhanced in the future.

Step 2: Provide Recommendations for Future Sustainability Enhancements
- Suggest **new features or design improvements** that could increase the product's sustainability in future versions.

- **Examples:**
  - Use **more sustainable packaging** made from compostable materials.
  - Transition to **carbon-neutral manufacturing** through renewable energy.
  - Implement a **take-back program** to recycle old products and minimize waste.

Step 3: Reflect on the Sustainability Design Process
- If you were to **restart the design process**, what would you change to improve sustainability?
- Summarize **lessons learned** from designing with sustainability in mind, including the importance of early planning and collaboration with suppliers.

Deliverable:
- A **Recommendations and Lessons Learned Report** (500-800 words) that:
  - Summarizes test results and feedback.
  - Provides actionable recommendations for future sustainability improvements.
  - Reflects on key takeaways from the design process.

Part 5: Compile the Final Sustainability Report

Step 1: Organize Your Report
- Compile all deliverables into a **Final Sustainability Report**, including:
  - Summary of sustainability considerations.
  - List of key sustainability metrics and how they were evaluated.
  - Documentation of sustainability features implemented.
  - Photos or diagrams of the product build.
  - List of omitted sustainability features with explanations.
  - Recommendations for future improvements and lessons learned.

Step 2: Submit Your Final Sustainability Report
- Ensure the report is **well-organized and professionally formatted**, with clear section headings for each part of the assignment.

Deliverable:
- A **PDF report** containing all required elements, submitted by the deadline.

Timeline and Suggested Schedule:
- **Week 1:** Identify sustainability considerations and define metrics.
- **Week 2:** Design and implement sustainability features in the product.
- **Week 3:** Evaluate sustainability performance and gather feedback.
- **Week 4:** Document findings, recommend improvements, and submit the final report.

Evaluation Criteria:
- **Completeness:** Were all required sections of the assignment completed and documented?
- **Quality of Sustainability Design:** Were meaningful sustainability features incorporated and tested?
- **Effectiveness of Testing:** Were sustainability features thoroughly evaluated?
- **Recommendations:** Were thoughtful recommendations provided for future sustainability improvements?
- **Professionalism:** Is the final report well-organized, clear, and polished?

18.10 Assignments

Assignment 18-9. Engineering ethics diagnostic quiz.

1. An engineer is tasked with designing a new automotive component. During the design process, they discover that the component may have a shorter lifespan than expected, potentially leading to increased maintenance costs for consumers. What should the engineer do?
    - A) Proceed with the design as planned, as the lifespan is not a critical safety issue.
    - B) Modify the design to improve the component's lifespan, even if it increases production costs.
    - C) Inform the project manager of the potential issue and let them decide how to proceed.
    - D) Conduct a cost-benefit analysis to determine the most appropriate course of action.

2. An engineer is working on a project that involves the design of a new medical device. During testing, they discover that the device has a small chance of causing an adverse reaction in patients with a rare genetic condition. What should the engineer do?
    - A) Proceed with the design as planned, as the risk is minimal and only affects a small population.
    - B) Modify the design to eliminate the risk of adverse reactions, even if it significantly increases development time and costs.
    - C) Inform the project team of the potential risk and collaborate to find a solution that balances patient safety and project constraints.
    - D) Halt the project and inform regulatory authorities of the potential risk.

3. An engineer is asked to design a new industrial machine that will significantly increase production efficiency. However, the machine may pose a safety risk to workers if not operated properly. What should the engineer do?
    - A) Design the machine as requested, as worker safety is the responsibility of the employer.
    - B) Refuse to design the machine due to the potential safety risk.
    - C) Design the machine with additional safety features and clear operating instructions to mitigate the risk.
    - D) Suggest an alternative design that prioritizes worker safety over production efficiency.

4. An engineer is working on a project that involves the design of a new consumer product. They discover that the product may have a negative environmental impact

during its lifecycle. What should the engineer do?
- A) Proceed with the design as planned, as environmental impact is not the primary concern.
- B) Modify the design to minimize the environmental impact, even if it increases production costs.
- C) Inform the project manager of the potential environmental impact and let them decide how to proceed.
- D) Propose an alternative design that prioritizes environmental sustainability.

5. An engineer is tasked with selecting a supplier for a critical component in a new product. They discover that the supplier with the lowest cost has a history of using child labor in their manufacturing facilities. What should the engineer do?
    - A) Select the supplier with the lowest cost to minimize project expenses.
    - B) Select a more expensive supplier that adheres to ethical labor practices.
    - C) Inform the project manager of the ethical concerns and let them make the final decision.
    - D) Propose a compromise solution that balances cost and ethical considerations.

6. An engineer is working on a project that involves the design of a new aerospace component. During the design process, they identify an opportunity to reduce the weight of the component, which would result in fuel savings for the end-user. However, the weight reduction may also slightly decrease the component's durability. What should the engineer do?
    - A) Prioritize weight reduction to maximize fuel savings, as it provides the greatest benefit to the end-user.
    - B) Maintain the original design to ensure maximum durability, even if it means forgoing potential fuel savings.
    - C) Conduct a risk assessment to determine the potential impact of the weight reduction on the component's performance and safety.
    - D) Consult with the project stakeholders to determine their priorities and make a decision based on their feedback.

18.10 Assignments

8. An engineer is asked to design a new software system that will be used to process sensitive personal data. They are concerned about the potential for the system to be hacked or misused. What should the engineer do?

    A) Design the system as requested, as data security is the responsibility of the end-user.

    B) Refuse to work on the project due to the potential risks associated with handling sensitive data.

    C) Design the system with robust security features and strict access controls to minimize the risk of data breaches or misuse.

    D) Suggest an alternative design that does not involve the processing of sensitive personal data.

9. An engineer is working on a project that involves the development of a new construction material. During testing, they discover that the material has the potential to release harmful chemicals if not disposed of properly. What should the engineer do?

    A) Proceed with the development as planned, as proper disposal is the responsibility of the end-user.

    B) Modify the material composition to eliminate the potential for harmful chemical release, even if it increases production costs.

    C) Provide clear disposal instructions and warnings to end-users to mitigate the risk of improper disposal.

    D) Collaborate with waste management experts to develop a comprehensive disposal and recycling plan for the material.

10. An engineer is tasked with designing a new recreational drone. During the design process, they identify an opportunity to increase the drone's range and speed, which would make it more appealing to consumers. However, the increased capabilities may also make the drone more difficult to control and potentially hazardous in inexperienced hands. What should the engineer do?

    A) Prioritize the drone's performance to maximize consumer appeal and market success.

    B) Limit the drone's capabilities to ensure safe operation, even if it reduces consumer appeal.

    C) Incorporate adjustable performance settings and comprehensive safety features to balance performance and safety.

    D) Consult with regulatory authorities to determine the appropriate balance of performance and safety features.

12. An engineer is working on a project that involves the design of a new wearable health monitoring device. During the development process, they become aware of a similar product being developed by a competitor. The engineer realizes that they could incorporate some of the competitor's innovative features into their own design, potentially saving time and development costs. What should the engineer do?
    A) Incorporate the competitor's features into the design without informing the project team, as it will benefit the project.
    B) Inform the project manager of the potential to incorporate the competitor's features and let them decide how to proceed.
    C) Refrain from using any of the competitor's features and focus on developing original solutions.
    D) Analyze the competitor's product to understand the underlying principles and use that knowledge to inform the development of original features.

13. An engineer is working on a project to design a new water treatment system for a developing country. They discover that the most effective solution would require the use of a patented technology, which would significantly increase the cost of the project. What should the engineer do?
    A) Use the patented technology without permission to maximize the effectiveness of the water treatment system.
    B) Seek alternative solutions that do not infringe on the patent, even if they are less effective.
    C) Negotiate with the patent holder to obtain permission to use the technology at a reduced cost for the humanitarian project.
    D) Proceed with the project using the patented technology and pass the increased costs on to the end-users.

14. An engineer is tasked with designing a new theme park attraction. During the design process, they identify an opportunity to reduce the construction costs by using lower-grade materials. The engineer believes that the lower-grade materials will still meet the minimum safety requirements. What should the engineer do?
    A) Use the lower-grade materials to reduce construction costs and maximize profit for the theme park.
    B) Use the higher-grade materials to ensure the highest level of safety, even if it increases construction costs.
    C) Conduct a thorough risk assessment to determine if the lower-grade materials provide an acceptable level of safety.
    D) Propose an alternative design that uses a combination of lower-grade and higher-grade materials to balance cost and safety.

18.10 Assignments

15. An engineer is working on a project to design a new type of prosthetic limb. They discover that the most advanced technology for the prosthetic is extremely expensive, which would make the final product unaffordable for many potential users. What should the engineer do?
    A) Use the most advanced technology to create the best possible prosthetic, regardless of the final cost.
    B) Use lower-cost components to make the prosthetic more affordable, even if it compromises performance.
    C) Develop a modular design that allows users to choose between basic and advanced components based on their needs and budget.
    D) Focus on reducing the cost of the advanced technology through innovative manufacturing processes and materials.

16. An engineer is working on a project to design a new type of agricultural drone. During the development process, they become aware that the drone could potentially be used for military purposes. What should the engineer do?
    A) Proceed with the design as planned, as the potential military use is not the engineer's responsibility.
    B) Modify the design to include safeguards that prevent the drone from being used for military purposes.
    C) Refuse to continue working on the project due to the potential for military use.
    D) Consult with the project stakeholders to establish clear guidelines and restrictions on the drone's use.

17. An engineer is tasked with designing a new social media platform. During the development process, they become concerned about the potential for the platform to be used to spread misinformation and hate speech. What should the engineer do?
    A) Proceed with the design as planned, as the content on the platform is not the engineer's responsibility.
    B) Incorporate robust content moderation and fact-checking systems to minimize the spread of misinformation and hate speech.
    C) Refuse to work on the project due to the potential for misuse of the platform.
    D) Propose an alternative design that prioritizes user privacy and limits the ability for content to go viral.

19. An engineer is working on a project to design a new type of autonomous vehicle. During testing, they discover a scenario in which the vehicle's decision-making algorithm may prioritize the safety of the passengers over the safety of pedestrians. What should the engineer do?
    A) Proceed with the design as planned, as the safety of the passengers should be the top priority.
    B) Modify the algorithm to prioritize the safety of pedestrians over the safety of passengers in all scenarios.
    C) Develop a more nuanced decision-making framework that balances the safety of all parties involved and minimizes overall harm.
    D) Refuse to work on the project due to the ethical dilemma posed by the autonomous vehicle's decision-making.

20. An engineer is tasked with designing a new type of security system for a government building. During the development process, they become concerned about the potential for the system to be used to violate the privacy rights of individuals. What should the engineer do?
    A) Proceed with the design as planned, as the use of the system is not the engineer's responsibility.
    B) Incorporate safeguards and strict access controls to minimize the potential for misuse of the system.
    C) Refuse to work on the project due to the potential for violation of privacy rights.
    D) Propose an alternative design that focuses on physical security measures rather than surveillance technology.

21. An engineer is working on a project to design a new type of renewable energy system. During the development process, they discover that the most efficient design would require the displacement of a small local community. What should the engineer do?
    A) Proceed with the most efficient design, as the benefits of renewable energy outweigh the impact on the local community.
    B) Modify the design to avoid displacing the local community, even if it reduces the efficiency of the energy system.
    C) Consult with the local community to understand their concerns and develop a mutually beneficial solution.
    D) Propose an alternative location for the renewable energy system that does not impact any local communities.

22. An engineer is tasked with designing a new type of medical device that uses artificial intelligence to assist with diagnosis and treatment decisions. During testing, they discover that the AI system may exhibit bias against certain demographic groups. What should the engineer do?
    A) Proceed with the design as planned, as the potential for bias is an inherent risk of using AI technology.
    B) Modify the AI system to eliminate any potential for bias, even if it reduces the overall accuracy of the device.
    C) Develop a rigorous testing and monitoring process to identify and mitigate any instances of bias in the AI system.
    D) Refuse to work on the project due to the ethical concerns surrounding the use of AI in medical decision-making.

23. An engineer is working on a project to design a new type of consumer electronics device. During the development process, they discover that the device could be easily modified by users to circumvent copyright protection measures on digital media. What should the engineer do?
    A) Proceed with the design as planned, as the potential for misuse is not the engineer's responsibility.
    B) Modify the device to prevent any possibility of circumventing copyright protection measures.
    C) Include clear warnings and disclaimers about the legal consequences of modifying the device to circumvent copyright protection.
    D) Consult with legal experts to ensure that the device complies with all relevant copyright laws and regulations.

24. An engineer discovers a serious safety flaw in a product that has already been released to the market. What is the most ethical course of action?
    A) Immediately inform the company and recommend a product recall.
    B) Discuss the issue with colleagues but take no further action.
    C) Resign from the company to avoid association with the defective product.
    D) Wait until there are reported incidents before taking action.

26. An engineering firm is competing for a large government contract. During the bidding process, a government official hints that a substantial donation to their reelection campaign could influence the decision. What is the most ethical response?
    A) Make the donation to secure the contract for the firm.
    B) Ignore the hint and proceed with the standard bidding process.
    C) Withdraw from the bidding process and report the official's behavior through proper channels.
    D) Discuss the situation with the official to understand their expectations better.

27. An engineer is asked to review and approve a design that was created by a colleague. The engineer notices several errors that could lead to structural failures if not addressed. When confronted, the colleague becomes defensive and dismissive. What should the engineer do?
    A) Approve the design to avoid damaging the colleague's reputation.
    B) Discuss the errors with the colleague and try to reach a resolution.
    C) Escalate the issue to a supervisor and recommend not approving the design until the errors are fixed.
    D) Make the necessary changes without informing the colleague.

28. An engineer is offered a significant financial incentive by a contractor to specify their products in a design, even though they may not be the best choice for the project. What is the most ethical course of action?
    A) Accept the incentive and specify the contractor's products.
    B) Reject the incentive and make an objective decision based on project requirements.
    C) Accept the incentive but donate it to charity to avoid personal gain.
    D) Specify the contractor's products but disclose the incentive to the client.

29. An engineer discovers that a colleague has been falsifying test results to meet project deadlines. What is the most appropriate response?
    A) Confront the colleague and demand they correct the results.
    B) Report the misconduct to the appropriate authorities within the organization.
    C) Ignore the issue to avoid damaging the colleague's career.
    D) Discuss the situation with other colleagues to gauge their opinions.

31. An engineer is working on a project that could have negative environmental impacts. The client pressures the engineer to downplay these impacts in the environmental assessment report. What should the engineer do?

   A) Comply with the client's request to maintain a good relationship.
   B) Refuse to modify the report and accurately present the potential environmental impacts.
   C) Modify the report but include a disclaimer about the client's influence.
   D) Remove the engineer's name from the report to avoid association with the modified content.

32. An engineer is involved in the design of a new product that will be manufactured in a developing country with lax labor laws. The engineer becomes aware that the manufacturing process may exploit workers. What is the most ethical course of action?

   A) Proceed with the design as planned, as labor practices are not the engineer's responsibility.
   B) Raise concerns about the labor practices to company leadership and advocate for fair treatment of workers.
   C) Resign from the project to avoid contributing to unethical labor practices.
   D) Discuss the issue with the workers directly and encourage them to advocate for better conditions.

33. An engineer is asked to design a product that will have planned obsolescence, meaning it will be designed to fail after a certain period to encourage repeat sales. What is the most ethical response?

   A) Design the product as requested, as planned obsolescence is a common business strategy.
   B) Refuse to participate in the project due to the unethical nature of planned obsolescence.
   C) Design the product with planned obsolescence but inform customers of the expected lifespan.
   D) Propose an alternative design that focuses on durability and longevity.

35. An engineer discovers that a supplier has been using substandard materials in a construction project. The supplier offers the engineer a significant bribe to overlook the issue. What should the engineer do?
    A) Accept the bribe and continue using the substandard materials to save costs.
    B) Reject the bribe and report the supplier's unethical behavior to the appropriate authorities.
    C) Negotiate a higher bribe to compensate for the risk of using substandard materials.
    D) Ignore the issue and allow the construction project to continue as planned.

36. An engineer is part of a team developing a new software system that collects and analyzes user data. The engineer becomes concerned about the potential misuse of this data and the lack of privacy safeguards. What is the most ethical course of action?
    A) Ignore the concerns and continue development as planned.
    B) Discuss the privacy concerns with the team and advocate for implementing stronger safeguards.
    C) Leak information about the project to the media to raise public awareness.
    D) Resign from the project without explaining the reasons to avoid conflict.

19 Documentation of Capstone Design Projects

19.1 Introduction

Documentation in engineering design encompasses a broad scope and a diverse range of subject areas. It integrates writing, organizing, project management, critical thinking, analysis, and engineering problem-solving. Adequate documentation is governed by a set of rules, guidelines, and best practices that ensure clarity, consistency, and comprehensiveness. Proper documentation is crucial for capturing lessons learned, facilitating knowledge transfer, and enhancing the success of design projects. It is also essential for meeting the expectations of sponsors and preparing students for professional engineering practices.

19.1.1 Importance of Formal Documentation

A structured approach to documentation significantly increases the probability of a design project's success. Without formal guidelines, essential details and lessons from past designs may be lost, reducing the efficacy of future projects. Establishing clear rules for documentation in capstone design classes is vital for several reasons:

1. **Knowledge Retention**: Captures lessons learned and best practices for future reference.
2. **Learning by Doing**: Enhances students' skills in documentation, benefiting their professional careers.
3. **Sponsor Satisfaction**: Meets sponsor expectations through comprehensive design reports, presentations, and regular updates.

4. **Process Clarity**: Ensures all stakeholders can follow and understand the design process and decisions made.

19.1.2 Engineering Design Documentation

Documentation is a critical element of the engineering design process. It captures the complexity of engineering design, involving numerous analyses and decision-making points based on data and evidence. Adequate documentation ensures that all aspects of the design process are recorded, allowing others to follow the rationale behind design decisions and the final results.

19.1.2.1 Key Components of Documentation

1. **Capstone Design Workbook**
   - **Purpose**: Acts as a diary of the project, recording the evolution of ideas, problems encountered, drawings, schematics, readings, and assignments.
   - **Content**: Includes detailed information, readings, assignments, research notes, lecture notes, design concepts, solutions considered, decisions made, and guidelines for design steps.
2. **Meeting Minutes**
   - **Purpose**: Capture discussions and incremental work performed during meetings.
   - **Content**: Document design steps, critical decisions, and action items. Ensure clarity and detail to facilitate follow-up and accountability.
3. **Design Presentation Slides**
   - **Purpose**: Communicate critical information at various stages of the project.
   - **Content**: Include details on design reviews, proof of concept, and testing phases. Used for stakeholder presentations to provide updates and gather feedback.
4. **Preliminary Design Report (PDR)**
   - **Purpose**: Capture the design process details leading up to the proof of concept and prototype.
   - **Content**: Includes background research, design requirements, initial concepts, and development steps. Serves as a milestone document for sponsor review and approval.
5. **Brochures and Posters**
   - **Purpose**: Summarize the project for a broader audience, such as conferences or design showcases.
   - **Content**: Highlight critical aspects of the project, including objectives, methodologies, results, and potential impacts.
6. **Final Design Report**

- o **Purpose**: Comprehensive documentation of the entire design project.
- o **Content**: Captures all critical aspects of the design work, including final designs, detailed analysis, test results, and lessons learned. Serves as the definitive record of the project for sponsors and future reference. Additionally, these reports serve as a part of an engineering student portfolio. During job interviews, capstone design projects often become the main focus, demonstrating the student's competency and ability to complete a major engineering project while working in a team.

19.1.3 Guidelines for Effective Documentation

1. **Consistency**: Use a uniform format and structure for all documents to ensure ease of reading and comprehension.
2. **Clarity**: Write clearly and concisely, avoiding jargon and technical language that all stakeholders may not understand.
3. **Detail**: Include sufficient detail to allow others to understand the design process, decisions made, and outcomes achieved.
4. **Accessibility**: Organize documents in a manner that allows easy access and retrieval of information.
5. **Regular Updates**: Maintain documentation regularly throughout the project to capture real-time progress and changes.

Adequate documentation is integral to the success of engineering design projects. It ensures that the design process is transparent, decisions are well-documented, and knowledge is preserved for future use. By adhering to established guidelines and maintaining thorough documentation, design teams can enhance project outcomes, meet sponsor expectations, and build valuable skills for their professional careers. Moreover, the capstone design reports serve as a valuable portfolio component, showcasing students' abilities and achievements to potential employers and affirming their competence in executing major engineering projects.

19.1.4 Design Binder Implemented as a Shared Electronic Drive

A design binder, traditionally a physical three-ring binder, can be efficiently implemented as a shared electronic drive. This method allows the project team to collect, organize, and share their project materials seamlessly. For a typical engineering capstone project conducted over two terms, an electronic design binder provides easy access to all relevant documents and facilitates collaboration among team members, professors, mentors, and sponsors.

19.1.4.1 Benefits of a Shared Electronic Design Binder:

1. **Accessibility**: Team members can access the binder from any location, promoting collaboration and real-time updates.
2. **Organization**: Digital folders and subfolders help keep materials well-organized and easy to navigate.
3. **Security**: Cloud storage services offer secure backups and version control, preventing data loss.

19.1.4.2 Key Components and Organization:

1. **Index and Table of Contents**
    - **Purpose**: Provide a roadmap of the binder's content.
    - **Content**: Include headings for each section.
2. **Problem Definition**
    - **Purpose**: Detail the problem being addressed.
    - **Content**: Include supporting materials created by the team.
3. **Team Work**
    - **Purpose**: Document team interactions and progress.
    - **Content**: Include meeting minutes, progress reports, and email correspondence with sponsors.
4. **Engineering Analysis**
    - **Purpose**: Capture technical details and analyses.
    - **Content**: Include notes, sketches, calculations, drawings, charts, and schematics.
5. **Project Plan**
    - **Purpose**: Outline the project timeline and tasks.
    - **Content**: Include baseline and updated project plans, Gantt charts, and task schedules with milestones.
6. **Presentations**
    - **Purpose**: Archive all project presentations.
    - **Content**: Include PowerPoint slides from all stages of the project.
7. **Literature and Patent Searches**
    - **Purpose**: Record background research.
    - **Content**: Include articles, book sections, journal articles, and patents.
8. **Design Approaches**
    - **Purpose**: Document design methods used.
    - **Content**: Include descriptions of methods and their applications to the project.
9. **Design Specifications**
    - **Purpose**: Record detailed design specifications.
    - **Content**: Include the revision history and dates.
10. **BOM (Bill of Materials)**

19.1 Introduction

- o **Purpose**: Track materials and suppliers.
- o **Content**: Include a cross-referenced BOM with supplier and contact lists.

11. **Systems Analysis**
    - o **Purpose**: Capture system-level analyses.
    - o **Content**: Include computational analyses, such as load-bearing capacity, strength, heat transfer, and materials behavior.
12. **References**
    - o **Purpose**: Record all external references.
    - o **Content**: Include competition rules, industry materials, standards, papers, and manuals.
13. **Modeling**
    - o **Purpose**: Archive all modeling efforts.
    - o **Content**: Include models and simulations using tools like CAD, Abaqus, Fluent, and Comsol.
14. **Trade-Off Analysis**
    - o **Purpose**: Document decision-making processes.
    - o **Content**: Include radar/spider charts, QFD models, and formal decision models.
15. **Financial Analysis**
    - o **Purpose**: Record cost analyses.
    - o **Content**: Include cost estimates, personnel effort, manufacturing costs, and facility usage.
16. **Critical Thinking and Analysis**
    - o **Purpose**: Capture broader analysis results.
    - o **Content**: Include analyses of economic, environmental, social, political, ethical, health and safety, manufacturability, and sustainability factors.
17. **Administrative**
    - o **Purpose**: Track administrative tasks.
    - o **Content**: Include purchase orders, bid sheets, quotations, registrations, and other paperwork.
18. **Resumes**
    - o **Purpose**: Record team member resumes.
    - o **Content**: Include the most recent resumes for all team members.

19.1.4.3   Practical Style Guide for Digital Organization:

- **Use a well-structured folder hierarchy**: Organize your documents in folders by sections similar to the traditional binder format.
- **Use consistent naming conventions**: Ensure all files and folders follow a consistent naming scheme for easy retrieval.
- **Regularly update and backup**: To prevent data loss, keep the electronic design

binder up-to-date and regularly backed up.

19.1.5 Electronic Files and Project Archive

Electronic files created during the design project are an essential part of the design work's documentation. The electronic files for the design project should be organized and available to all team members (including professors and teaching assistants) and possibly mentors and sponsors. A suggested organization of information is presented in the following list of folder names:

- Additional considerations
- Administrative
- Assessment
- Brochures
- CAD Files
- Concepts
- Cost Analysis
- Critical Design Review
- Design for X
- Design Specifications
- Final Design Report
- Manuals
- Meeting Minutes and Notes
- Paper
- Patent Search
- Photos
- Poster
- Preliminary Design Report
- Presentations
- Previous Project Information
- Problem Definition
- Project Management
- QFD Analysis
- References
- Resumes
- Testing
- Videos
- Weekly Progress Reports

Electronic files can be shared with cloud file systems such as Google Drive or Dropbox. Many universities have standardized the Google application suite (G Suite), which facilitates collaboration and document sharing.

Adequate documentation is integral to the success of engineering design projects. It ensures that the design process is transparent, decisions are well-documented, and knowledge is preserved for future use. By adhering to established guidelines and maintaining thorough documentation, design teams can enhance project outcomes, meet sponsor expectations, and build valuable skills for their professional careers. Moreover, the capstone design reports serve as a valuable portfolio component, showcasing students' abilities and achievements to potential employers and affirming their competence in executing major engineering projects.

19.2 Verbal Presentation with Slides

Verbal team presentations are a cornerstone of capstone design projects. Typically, student design teams are expected to make two or three presentations over an academic term. These presentations, which range from 10 to 30 minutes, must be meticulously prepared and delivered to communicate progress, accomplishments, and future directions effectively. To achieve a professional presentation, teams must practice extensively and adhere to specific guidelines.

19.2.1 Tips for Preparing an Excellent Presentation

1. **Plan Your Presentation Structure**
    - Ensure it includes an introduction, key points, and a conclusion.
    - Use a logical flow to guide the audience through your content.

2. **Enhance Visual Communication**
    - Use PowerPoint to organize and present complex information clearly.
    - Select a clean, professional template with consistent formatting.

3. **Use a Cohesive Color Scheme**
    - Ensure text and visuals are well-aligned.
    - Stick to a cohesive color scheme that enhances readability.

4. **Follow the 6x6 Rule**
    - Use no more than six bullet points per slide.
    - Each bullet should have no more than six words.

5. **Keep Content Concise**

- Avoid overloading slides with information.
- Make sure every slide serves a purpose and avoid redundancy.

6. **Control Your Timing**
   - Allocate 1-2 minutes per slide.
   - Practice to ensure you stay within the allotted time.

7. **Speak Clearly and Confidently**
   - Make eye contact with the audience.
   - Use clear and descriptive titles for each slide.

8. **Tailor to Your Audience**
   - Ensure your content is relevant and engaging.
   - State the purpose of your presentation clearly at the beginning.

9. **Tell a Compelling Story**
   - Highlight the journey from problem definition to solution development.
   - Focus on critical points and significant findings without getting lost in details.

10. **Structure Logically**
    - Use transitions to connect different sections.
    - Aim for a natural and engaging delivery with controlled pacing.

11. **Ensure Accessibility**
    - Use high-contrast colors and large fonts.
    - Test readability in different lighting conditions.

12. **Use Graphics Effectively**
    - Include relevant, high-quality graphics to illustrate key points.
    - Avoid overly bright or clashing colors and ensure good contrast.

13. **Make It Enjoyable**
    - Use humor or interesting anecdotes where appropriate.
    - Keep the audience engaged with interactive elements or questions.

14. **Refocus Attention**
    - Periodically summarize key sections before moving on.
    - Recap the main takeaways of your presentation at the end.

15. **Proofread and Edit**
    - Carefully proofread all slides for clarity, accuracy, and conciseness.
    - Avoid common PowerPoint pitfalls like overloading slides with text.

16. **Utilize PowerPoint Features**
    - Use transitions, animations, and multimedia effectively.
    - Ensure they enhance, rather than detract from, your message.

17. **Prepare for Q&A**
    - Allocate time for a Q&A session at the end.
    - Prepare for potential questions and answer them confidently.

19.2.2 Guidelines for Capstone Design Presentations

Effective design presentations require adherence to specific guidelines that ensure all team members participate and the presentation is professional, efficient, and informative. Given the limited time available due to class size and scheduled meeting hours, the following guidelines can help teams prepare:

1. **Time Management**
    - Establish a maximum time limit for each presentation (e.g., 15, 20, 25, or 30 minutes), including setup, questions, and answers.
    - Aim to keep the presentation to 80% of the allotted time to allow for setup and Q&A.

2. **Inviting Sponsors**
    - Ensure sponsors are invited to the presentations.
    - Inform your professor if sponsors will attend.

3. **Team Participation**
    - All team members must participate in the presentation.
    - Assign specific slides to each member to practice and present.

4. **Note-taking**

- Assign a team member to take notes on questions, comments, and suggestions from the audience.

5. **Introductions**
    - Introduce all team members and their roles at the beginning of the presentation.

6. **Presentation Content**
    - Include team organization and division of responsibilities.
    - Define the problem and provide a design specifications table.
    - Present design concepts and how they address the sponsor's problem.
    - Conduct engineering analysis, testing, validation, and verification using relevant tools and software.
    - Include Pugh and QFD analysis when appropriate.
    - Conduct a trade-off analysis of alternatives.
    - List problems and unresolved issues and explain plans to address them.
    - Summarize and conclude the presentation, highlighting creativity and innovation.

7. **Preparation and Submission**
    - Provide a copy of the presentation (e.g., PowerPoint file) to the professor and sponsors before the presentation.

8. **Assessment**
    - The presentation will be evaluated on the team's ability to work together, the effectiveness of the presentation, and the probability of achieving a successful design solution.

9. **Audience Participation**
    - Encourage the audience (students, professors, mentors, and sponsors) to provide meaningful critiques and comments.

10. **Positive Attitude**
    - Maintain a positive attitude throughout the presentation, demonstrating enthusiasm and confidence in the project.

By following these guidelines and tips, capstone design teams can deliver effective, professional presentations that clearly communicate their project's progress and future direction.

19.3 Photos

Photographs have been utilized as sources of evidence and information in scientific endeavors for over a century. In recent years, the camera has become ubiquitous, integrated into virtually every cell phone, making it an ever-available tool to capture observations instantly. In the context of engineering, and particularly in capstone design projects, photographs are invaluable for documenting the progression of design work, capturing precise details that written descriptions alone may not fully convey.

19.3.1 Importance of Photographs in Engineering Design

1. **Precision and Clarity**
   - A photograph can convey complex design details more precisely than many words. It provides an exact likeness of the objects or processes captured, reducing ambiguity and enhancing understanding.
   - Visual documentation ensures that the exact features, dimensions, and conditions of the design are accurately recorded and easily interpreted by anyone reviewing the project.
2. **Persuasive Evidence**
   - Photographs serve as persuasive evidence, showing the actual state of a design at various stages. This visual record is compelling because it provides concrete proof of what was achieved and how it was done.
   - Unlike text descriptions that rely on the reader's imagination and interpretation, photographs present an objective reality that can be universally understood.
3. **Enhanced Understanding with Annotations**
   - Adding descriptive words and annotations to photographs further enhances comprehension. Labels, arrows, and notes can highlight critical components, explain functions, and illustrate procedures, making complex concepts accessible and transparent.
   - Annotated photographs can be used in presentations, reports, and documentation to provide detailed visual explanations that support and clarify written content.
4. **Documenting Design Evolution**
   - Capturing photographs at each stage of the design evolution, including assembly, parts, operation, maintenance, and fitting, provides a chronological record of the project's development.
   - This visual timeline helps track progress, identify changes, and reflect on the design decisions made throughout the project.
5. **Human Element**
   - Including photographs of design team members working on the

project adds value by documenting their participation and contributions. These images serve as historical records of who was involved, what roles they played, and where and when specific tasks were performed.
- People in photographs also provide a sense of scale and context, helping to understand the form, size, and function of the design elements in relation to human interaction.

19.3.2 Practical Applications of Photographs in Capstone Design Projects

1. **Capturing Design Stages**
   - Take photographs during critical stages of the design process to document initial concepts, prototypes, testing phases, and final products.
   - Use these images to illustrate the development and refinement of the design, showing the progression from idea to implementation.
2. **Recording Assembly and Maintenance**
   - Photograph the assembly process to document how components fit together, identify potential issues, and provide a visual guide for future assembly.
   - Capture images of maintenance procedures to create instructional materials that explain how to service and repair the design.
3. **Showing Operations and Functionality**
   - Use photographs to demonstrate how the design operates in real-world conditions. Highlight key features and mechanisms and show the design in action.
   - These images can be used in presentations and reports to provide a clear understanding of the design's functionality and performance.
4. **Including Annotations and Descriptions**
   - Annotate photographs with labels, arrows, and notes to explain specific features, parts, and functions. This helps clarify the images and provides detailed visual explanations.
   - Combine photographs with descriptive text to create comprehensive visual documentation that supports the written narrative of the project.

19.3.3 Photo Example

In a project focused on improving a magnetic seal demagnetizer, photographs can capture each stage of the design and testing process. Annotated images can highlight the components used, the setup of the testing apparatus, and the results observed. Including team members in these photographs adds context and demonstrates their active

19.3 Photos

involvement in the project.

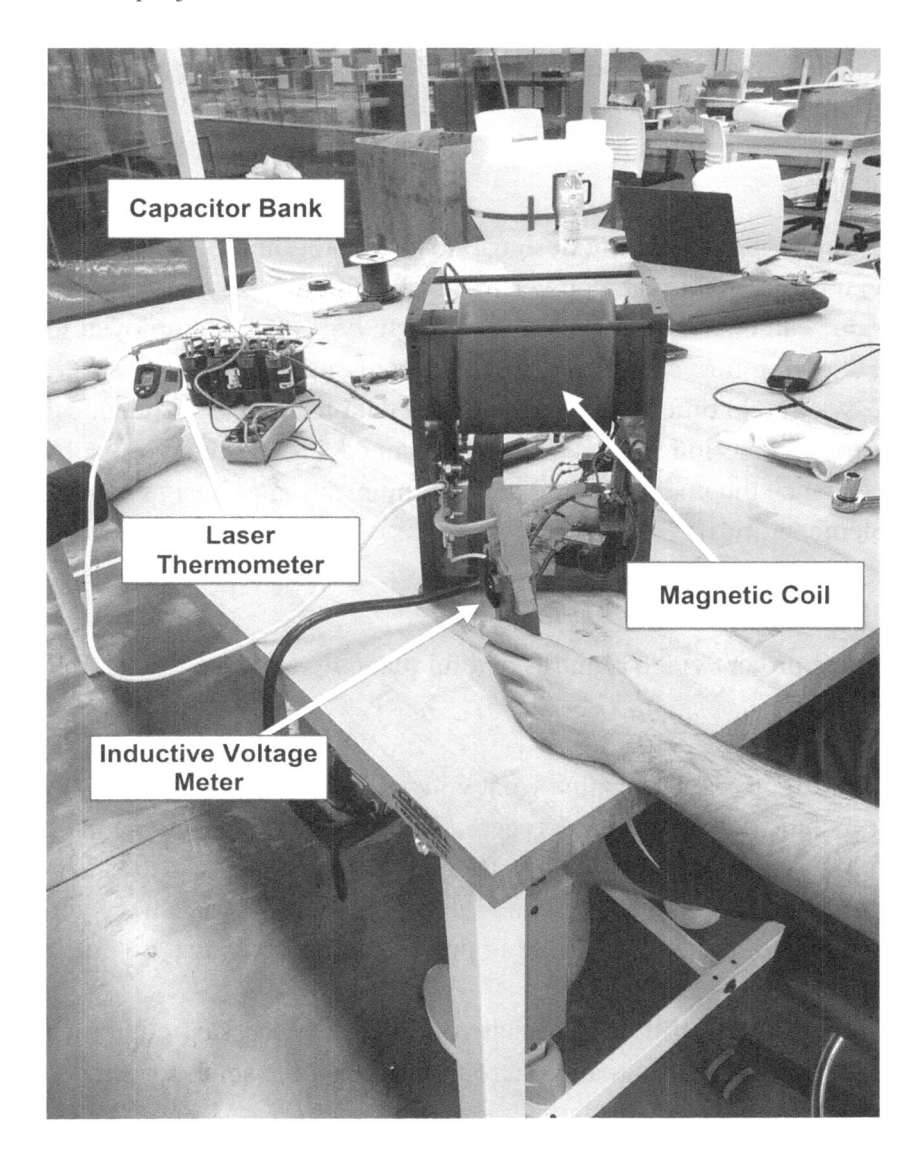

Figure 19-1. Example photograph showing magnetic seal demagnetizer.

Example Photograph Descriptions:
- **Initial Design Concept**: Photograph of the initial sketches and 3D models.
- **Prototype Assembly**: Images showing the step-by-step assembly of the prototype.
- **Testing Setup**: Photographs of the testing equipment and the prototype during testing, with annotations explaining the setup and measurements.
- **Team Involvement**: Pictures of team members working on different aspects of the project, such as assembling components, conducting tests, and analyzing results.

By systematically capturing and annotating photographs throughout the capstone design project, teams can create rich visual documentation that not only records their progress and achievements but also provides a clear and compelling narrative of their engineering journey.

19.4 Videography

Videography in engineering design extends photography's capabilities by adding the dimension of time, making it a powerful tool for capturing dynamic processes and events. The same cameras used for photography can typically be employed for videography, allowing for versatile documentation of engineering projects. Videos can capture processes in real time, providing valuable data for time-dependent calculations, process analysis, and motion studies. Properly planned and executed project videos can significantly enhance the understanding and communication of design concepts and their practical implementations.

19.4.1 Planning and Goals for Project Videos

Creating a project video requires careful planning and clear goals. Before filming, the team should:

1. **Define Objectives**: Determine what you want to capture and why. Objectives might include documenting a process, demonstrating a prototype in action, or recording user interactions with a design.
2. **Storyboard**: Outline the key scenes and shots required to achieve your objectives. This helps in organizing the filming process and ensures that no critical aspect is overlooked.
3. **Equipment and Setup**: Ensure that you have the necessary equipment, including cameras, tripods, lighting, and microphones. Plan the setup for each shot to capture clear and stable footage.
4. **Coordination**: If the video involves process operators or team members, coordinate with them to ensure smooth execution and minimal disruptions. Brief participants on their roles and the timing of their actions.

19.4.2 Timing Studies

Videography is particularly useful for time and motion analysis in process design and improvement. Software tools can be employed to accelerate the procedures for capturing and analyzing the timing of events, saving hours of manual effort and providing accurate documentation.

1. **Software Tools for Analysis**
    - Use specialized software to analyze video footage and measure time intervals, motions, and actions within a process.

- These tools can automatically track and record the timing of each event, reducing the potential for human error and increasing precision.
2. **Process Documentation**
    - Coordinate videography with process operators to ensure that all events, activities, or operational steps are captured comprehensively.
    - Document the baseline state of the process before any design changes are made. This provides a reference for comparison and analysis.
3. **Benefits of Video Time and Motion Studies**
    - Videos accurately document and time tasks while isolating non-value-added work content.
    - Video-supported analysis creates a verifiable history of the current process state, which can be invaluable for process optimization and redesign.

19.4.3 Position, Velocity, and Acceleration Measurement

Videography is also practical for measuring dynamic parameters such as position, velocity, and acceleration. This is essential in various engineering applications where understanding motion is critical.

1. **Capturing Motion**
    - Use high-frame-rate cameras to capture fast-moving objects or processes. This allows for a detailed analysis of motion that might be missed at standard frame rates.
    - Ensure that the camera setup minimizes motion blur and captures clear, precise footage.
2. **Analyzing Motion Data**
    - Employ video analysis software to track objects frame-by-frame, calculating position changes over time.
    - Use the data to compute velocity (rate of change of position) and acceleration (rate of change of velocity). This can be visualized through graphs and charts for better understanding.
3. **Applications in Engineering Design**
    - Measure the performance of mechanical components under operational conditions.
    - Analyze the dynamics of robotic arms, conveyor systems, or any other automated machinery.
    - Validate simulation models by comparing them with real-world motion data captured through video.

19.4.4 Practical Considerations for Effective Videography

1. **Lighting and Clarity**
   - Ensure proper lighting to avoid shadows and enhance the visibility of details. Good lighting is crucial for capturing clear and usable footage.
   - Use additional lights if necessary, and adjust the camera settings to optimize exposure.
2. **Sound Quality**
   - If the video includes spoken commentary or requires capturing ambient sounds, use external microphones to ensure high-quality audio.
   - Minimize background noise and ensure that speakers are close to the microphone for clear sound recording.
3. **Editing and Presentation**
   - Edit the video to remove unnecessary parts, enhance clarity, and add annotations or captions where needed.
   - Use editing software to splice together different clips, add transitions, and incorporate background music or voiceovers to improve the overall presentation.
4. **Storage and Sharing**
   - Store video files in organized folders with clear labels and descriptions. Use a structured naming convention for easy retrieval.
   - Share videos through cloud platforms like Google Drive or Dropbox to ensure accessibility for all team members, mentors, and sponsors.

19.4.5 Example Applications of Videography in Capstone Design Projects

1. **Prototyping and Testing**
   - Document the assembly and testing phases of prototypes, capturing detailed footage of each step. This can be used to review and improve the design.
   - Record tests to demonstrate the functionality and performance of the design, providing clear evidence for reports and presentations.
2. **Process Improvements**
   - Conduct time and motion studies to identify inefficiencies in current processes. Use video analysis to support redesign efforts aimed at optimizing workflow and reducing waste.
   - Measure and document improvements achieved through design changes, using before-and-after footage for comparison.
3. **Educational and Training Purposes**
   - Create instructional videos that explain complex processes or demonstrate the operation of equipment. These can be used for

training new team members or for educational purposes in related courses.
- Use videos to share knowledge and best practices within the team and with future capstone design students.

By incorporating videography into the documentation and analysis processes, engineering design teams can gain deeper insights into their projects, improve communication of their work, and create a robust visual record that complements other forms of documentation. This multimedia approach enhances the overall quality and effectiveness of the capstone design experience.

19.5 Poster

A poster presentation is a widely used format at scientific and engineering conferences, offering a formal visual presentation of research or design projects. Many engineering conferences, such as those organized by the American Society for Engineering Education (ASEE), have special sections dedicated to student poster presentations. These events provide an excellent opportunity for students to showcase their work, engage with attendees, and receive feedback from peers and experts in the field.

19.5.1 Elements of a Design Poster

A well-designed poster contains brief text, charts, graphs, and other visual aids that effectively convey information about the project. The following elements are typically included:

1. **Title and Author Information**
   - Title of the project
   - Names of team members
   - Institutional affiliation and contact information
2. **Abstract**
   - A concise summary of the project, including objectives, methods, results, and conclusions.
3. **Introduction**
   - Background information and context for the project.
   - The problem statement and the significance of the work.
4. **Objectives**
   - Clear and concise goals of the project.
5. **Methodology**
   - Description of the design process, including any theoretical or experimental methods used.
   - Flowcharts, diagrams, and process descriptions.

6. **Results**
    - Key findings illustrated with charts, graphs, and images.
    - Data visualization to highlight important results and trends.
7. **Discussion**
    - Interpretation of results and their implications.
    - Comparison with existing solutions or previous work.
8. **Conclusion**
    - Summary of the project's outcomes.
    - Potential future work and improvements.
9. **Acknowledgments**
    - Recognition of any sponsors, mentors, or contributors.
10. **References**
    - Cited sources and relevant literature.

19.5.2 Presentation Format

The typical format for a poster presentation involves attendees viewing and reading the poster while the design team is present to discuss the project and answer questions. This interactive format allows for in-depth discussions and direct feedback.

19.5.2.1 Technical Specifications

1. **Size**
    - Posters can vary in size, with the actual dimensions specified by the conference. For ASEE and many other student conferences, a size of 24 inches by 36 inches is recommended.
    - Another common size is 36 inches by 50 inches, depending on the conference requirements.
2. **Material**
    - Posters are usually printed on paper or vinyl and affixed to a corkboard or mounted on a foam core poster board.
    - Foam core poster boards are available in various sizes, including the standard 24x36 inches, and provide a sturdy backing for the poster.
3. **Printing**
    - Mid-range large format printers available at many engineering programs can easily handle the recommended 24x36 size.
    - Ensure high resolution for clarity and readability of text and images.

Figure 19-2. Example poster.

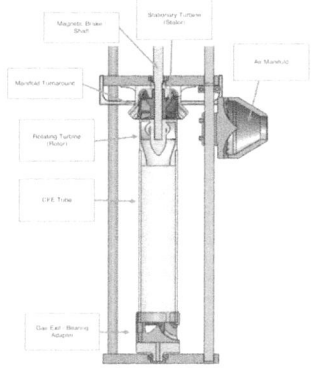

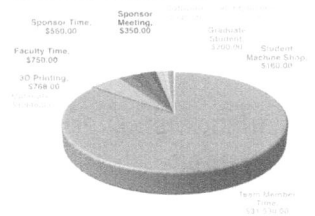

19.5.2.2 Tips for Creating an Effective Poster

1. **Visual Design**
   - Use a clean and professional layout with a logical flow of information.
   - Choose readable fonts and appropriate font sizes for different sections. Titles and headings should be larger, with body text in a smaller, readable size.
   - Employ color schemes that are visually appealing and enhance readability. Avoid using too many colors that can distract from the content.

2. **Content**
   - Keep text concise and to the point. Use bullet points and short paragraphs to make the information easily digestible.
   - Incorporate high-quality images, graphs, and charts to visually represent data and key points.
   - Use captions to explain images and charts clearly.

3. **Interaction**
   - Be prepared to engage with viewers. Practice summarizing your project succinctly and be ready to answer detailed questions.
   - Encourage questions and discussions to gain valuable feedback and insights.

4. **Proofreading**
   - Carefully proofread the poster for spelling and grammatical errors.
   - Ensure all data and information are accurate and sources are properly cited.

19.5.2.3 Benefits of Poster Presentations

1. **Visibility**
   - Poster presentations provide visibility for your project within the academic and professional community.
   - They offer a platform to showcase innovative work and gain recognition.

2. **Networking**
   - Engaging with attendees fosters networking opportunities with professionals, potential collaborators, and future employers.
   - It allows for the exchange of ideas and constructive feedback.

3. **Skill Development**
   - Creating and presenting a poster enhances skills in visual communication, public speaking, and critical thinking.
   - It helps in refining the ability to summarize and present complex

information effectively.
4. **Feedback and Improvement**
    o Direct interaction with viewers provides immediate feedback, which can be used to improve the project.
    o Insights gained from discussions can lead to new ideas and directions for future work.

In summary, poster presentations are a valuable tool for engineering students to present their design projects. They combine visual appeal with concise information to communicate complex ideas effectively. By adhering to technical specifications, focusing on visual design, and preparing for interactive discussions, students can create impactful posters that showcase their work and foster meaningful professional engagement.

19.6 Technical Information Sheet

Compressing details about a design project into a single-page information sheet is an efficient and professional way to quickly convey information. The Technical Information Sheet (TIS) provides essential facts about the product or process in a compact format, making it easy for technical professionals to understand the problem and the design solution. Information sheets play a crucial role in communicating with all stakeholders, ensuring they grasp the critical parts of the design solution swiftly and effectively.

19.6.1 Purpose and Importance of the Technical Information Sheet

The TIS is a one-page document, typically double-sided, that compiles key information and data about the product or process design. It includes critical details, facts, and figures presented visually through charts, drawings, schematics, and images. The primary purposes of a TIS are:

1. **Efficient Communication**: Quickly conveys essential information, saving stakeholders' time.
2. **Professional Presentation**: Provides a polished, concise overview suitable for professional meetings and conferences.
3. **Stakeholder Understanding**: Ensures all stakeholders, including sponsors, mentors, and peers, have a clear understanding of the project's key aspects.

19.6.2 Key Components of a Technical Information Sheet

A well-crafted TIS should contain the following elements:

1. **Design Problem Definition**
    o Briefly describe the problem the design aims to solve.

- Explain the motivation for addressing this problem.
2. **Design Solution Information**
    - Summarize the design solution, including key features and benefits.
    - Highlight how the solution addresses the problem.
3. **Statistics and Data**
    - Present quantitative data supporting the design, such as performance metrics, test results, and specifications.
    - Use tables, charts, and graphs for clarity.
4. **Design Specifications**
    - List the design specifications in a concise, tabular form.
    - Include relevant technical data and standards.
5. **Visual Aids**
    - Incorporate drawings, schematics, and images to illustrate the design.
    - Use visual elements to enhance understanding and retention of information.
6. **Project Overview**
    - Provide a brief overview of the design process, including key milestones and achievements.
    - Mention any alternative concepts considered and the rationale for the final design.
7. **Conclusions and Future Recommendations**
    - Summarize the main findings and results of the project.
    - Offer recommendations for future work or improvements.

19.6.3 Guidelines for Capstone Technical Information Sheets

Creating a Capstone Technical Information Sheet (CTIS) requires attention to detail and adherence to specific guidelines to ensure clarity and effectiveness:

1. **Length and Format**
    - Preferably a single page, double-sided. This format is suitable for professional contexts.
    - Ensure the information presented is brief and concise.
2. **Visual Appeal**
    - Design the CTIS to be visually appealing, using tables, charts, drawings, schematics, graphs, and bullet points.
    - Balance visuals with text to maintain readability and engagement.
3. **Content and Structure**
    - The CTIS should include a balance of visuals, words, graphics, and illustrations.
    - Use white spaces, bullet points, and bold headings to enhance

readability.

4. **Practical Benefits**
   - Saves time by providing precise information in a compact format.
   - Easy to read, with a mix of visual and textual elements to facilitate quick comprehension.

19.6.4 Style Guide for Capstone Technical Information Sheet

A style guide ensures consistency and clarity in the creation of CTIS. The following sections are suggested for inclusion:

1. **Project Title**
   - Use the project title as the CTIS title.
   - Include the university logo and sponsoring company logos for affiliation.
2. **Design Problem Description and Motivation**
   - Provide a brief and precise description of the problem and the motivation for solving it.
   - Set the stage for the remaining information.
3. **Design Specifications**
   - List the design specifications in a concise, quantitative tabular form.
   - Include relevant metrics and standards.
4. **Product or Process Design Process**
   - Briefly describe the steps taken by the design team to achieve the solution.
   - Mention any alternative concepts considered.
5. **Design Solution Achieved, Tested, and Redesigned**
   - Summarize the design solution and any testing results.
   - Present the final design solution, including details of redesign cycles and improvements.
6. **Conclusions and Recommendations for the Future**
   - Highlight significant features and results of the project.
   - Provide recommendations for future work, guiding stakeholders and sponsors on potential improvements.

19.6.5 Example Layout of a CTIS

Front Side:

1. **Header**
   - Project Title
   - University and Sponsor Logos

2. **Introduction**
    - Brief Project Overview
    - Problem Description and Motivation
3. **Design Solution**
    - Key Features and Benefits
    - Visuals: Drawings, Schematics, Images
4. **Data and Statistics**
    - Performance Metrics
    - Tables and Charts

Back Side:
1. **Design Process**
    - Overview of Steps Taken
    - Key Milestones and Achievements
2. **Testing and Results**
    - Summary of Testing Procedures
    - Results and Data Visualization
3. **Conclusions**
    - Main Findings
    - Visuals: Graphs and Charts
4. **Future Recommendations**
    - Suggestions for Improvement
    - Potential Next Steps
5. **Contact Information**
    - Team Members
    - Supervisor or Mentor Contact

By adhering to these guidelines and incorporating the suggested components, a CTIS can effectively communicate the essential aspects of a capstone design project. It provides a professional, concise, and visually appealing summary that stakeholders can quickly understand and appreciate.

19.6 Technical Information Sheet

Mechanical Engineering Capstone Design 2023-2024

RAM Works

Joel Aubin	Nicholas Caito	Nicholas Colavecchio	Nicholas DeFranzo
Fluid Dynamics Engineer	Design Engineer	Simulations Engineer	Nuclear/Thermal Engineer

NASA Centrifugal Nuclear Thermal Rocket
Team 20H – Gas Turbine Integration with Fuel Tube

Where ingenuity and invention converge

SUMMARY

The purpose of this project is to design and evaluate the flow of the hydrogen propellant-coolant as it is transferred throughout the entire centrifugal nuclear thermal rocket (CNTR). The hydrogen must act as both propellant and coolant for the rocket. Due to the centrifugal liquid uranium design of the rocket, temperatures can reach 5000K, requiring advanced cooling to ensure that each centrifugal fuel element (CFE) remains operational. If successful, the CNTR concept would allow ~1800 Isp for fast round trips to Mars.

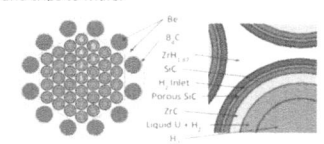

Figure 1. OSU Moderator Visual (above) and Measure (bottom)

METHODOLOGY

- **Simulation Strategy and Software Use:**
Due to the extreme conditions the rocket faces and the constraints in both resources and time, the team concentrated on using simulations to evaluate and refine designs. OpenFOAM was employed to simulate both the complete and specific sections of the flow path, aiding in design verification and overall improvement.
- **CAD Design and Revision:**
All parts were designed and iteratively refined using SolidWorks, ensuring each component met project specifications.
- **Mathematical Modeling:**
Development of a MATLAB code to analyze heat transfer and fluid dynamics within the rocket's geometry is ongoing, although it currently lacks models for specific flow characteristics needed for a complete analysis. The most relevant fluid dynamic model is the Taylor-Couette-Poiseuille Flow, shown below.

DESIGN

The property requirements for this CNTR concept are:

Property	Magnitude
Flow Pressure	~ 500 PSI
Flow Velocity	< Mach 1
Mass Flow Rate	0.2 kg/s
Specific Impulse	1800s
Heat Transfer Rate	Maximum Value Possible
Propellent Temperature	5000 K
Cylinder Rotation	7000 RPM

INTRODUCTION

NASA wishes to implement a centrifugal thermal power nuclear rocket to limit the adverse side-effects space imposes on astronauts by minimizing the transit time to travel to other planetary bodies. With this concept, the CNTR rocket demonstrated greater specific impulse, more power generation, and limited excess consumption of its fuel elements when compared to both solid and liquid fuel rockets. Team 20H is tasked with designing and implementing a hydrogen path-way system that meets the specific requirements elicited by NASA to fulfill the following purposes: Cool the moderator and shielding elements that are rapidly heated by the nuclear fuel elements, follow a turn-around geometry to cool the bearings allowing the fuel elements to spin, and exit the inner annulus region through a final nozzle as a powerful propellant.

TESTING AND REDESIGN

To compute the fluid properties along the complex geometries utilized in this project, computational fluid dynamics programs OpenFOAM and SIMSCALE were utilized to generate pressure and velocity readings which can be then utilized in MATLAB scripts that focus on interpolating viscous turbulent flow equations to obtain our other designated values. With the results obtained from the simulation, the CAD files of the simulated piece were redesigned to fix any areas of concern.

A major redesign was made to the modular system; previously, all centrifugal fuel elements (CFEs) shared one moderator block and one bottom pan. To address computational challenges and fluid flow issues, each CFE was redesigned to have its own hydrogen supply and bottom pan. This change simplifies simulations and enables the future integration of a control system to regulate hydrogen flow to each CFE independently.

Significant effort was invested in our physical model to minimize friction between components. The objectives were to transport gas from the reservoir to the outlet of the CFE and to determine if the turbine could be spun effectively. Initially, grease was used to reduce friction, but it proved insufficient. As a result, additional bearings were incorporated to facilitate smoother rotation. A depiction of this physical model is shown in the figure below.

Organized and taught by Professor B. Nassersharif, E-mail: bn@uri.edu, Phone: 401-874-9335

Figure 19-3. Example CTIS front side.

Mechanical Engineering Capstone Design 2023-2024 — RAM Works

SIMULATIONS

The primary source of testing designs involved OpenFOAM CFD simulations. Each part underwent various simulations with multiple different boundary conditions to determine the effectiveness of the design. The design simulations included full-path simulations, where the flow begun at the inlets of the moderator block and exited the rocket to open space to simulate the real flow path during operation. Similarly, partial-flow simulations were conducted on certain designs to get more accurate results around areas of concerns. Generally, these simulations included either full geometry or 1/6th symmetric geometry depending on the accuracy and computational intensity of the simulation.

The bottom shuriken design was simulated multiple times with various boundary conditions and design changes to improve its effectiveness and efficiency at distributing hydrogen. With the new modular design, only one CFE had to be simulated to understand the flow behavior. The shuriken bottom plate included a partial-path, full-geometry simulation to analyze the fluid as it turn 180 degrees into the annulus region from the inlet holes. The most recent simulations included a 15psi inlet boundary condition for each of the 6 inlet holes and an outlet condition of 0 Pa (exhaust to open space), shown below in Figure 2.

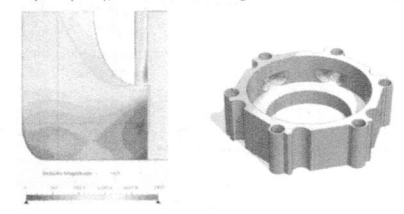

Figure 2. Shuriken, Full geometry, partial-path simulation, 15 psi inlet, 0 Pa outlet (left). CAD files of Shuriken (right)

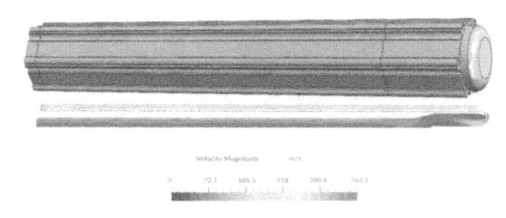

Full Single Modular CFE (Seen above)
- Full-path, 1/6th symmetric geometry, 1 inlet
- 9E-4 kg/s inlet, 0 Pa outlet
- 7000 RPM rotating inner annulus face (CFE)
- Results showed promising improvements over previous simulations, with the shuriken effectively delivering hydrogen to the annulus and the top turnaround, thereby increasing flow pressure for the turbine as shown below in Figure 4.

Figure 4. Top turnaround of full-path simulation, showing the increase in pressure before entering turbine

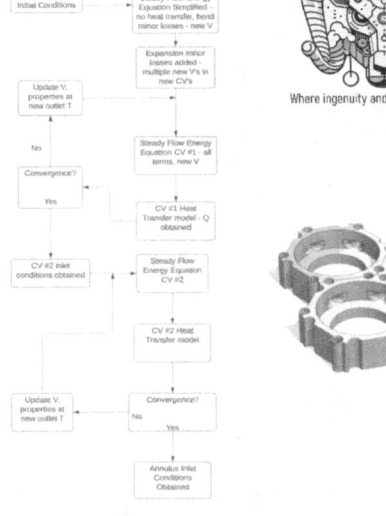

The MATLAB code analyzes fluid flow and heat transfer from the CNTR core's fluid channels through the moderator block to the annulus inlet. It employs the Steady Flow Energy Equation (SFEE) to estimate initial fluid velocities, which are then adjusted based on heat transfer models tailored to specific control volume geometries. The process iteratively updates velocity and heat transfer until values converge. These converged outlet values are then used as inlet values for the next control volume, continuing until the annulus is reached. Future updates will include Reynolds and Nusselt number correlations for Taylor-Couette-Poiseuille flow, necessitating adjustments to the iteration process and control volume models to accommodate co-dependent thermal-fluid analysis. As a result, the state diagram will be revised to incorporate these changes.

FURTHER WORK

- Finalizing the MATLAB script to obtain fluid conditions
- Characterizing the thermal profile of the liquid uranium
- Integrating both heat and fluid dynamics in one COMSOL simulation
- Improving upon and properly simulating the manifold design
- Designing an attachment between the physical model and a high-pressure source of gas to better direct the flow.

CONCLUSION

With the combined team effort, a physical model was created to best demonstrate the centermost CFE, which generates the highest temperature profile for the entire reactor. Each fluid simulation utilized a singular CFE in which the fluid properties obtained could be interpolated for the rest of the CFEs. After our simulations, it is concluded that the current design doesn't depict any major flaws including choked flow points, adverse pressure differential drops in non-designated areas, or trapped fluid flow regions in the turn-around region. Additionally, it can be concluded that the main purpose of this project, which is to create a gas integration system, has been accomplished.

Organized and taught by Professor B. Nassersharif, E-mail: bn@uri.edu, Phone: 401-874-9335

Figure 19-4. Example CTIS back side.

19.7 Final Design Report

The Final Design Report (FDR) is a comprehensive formal engineering document that captures the entire design project's accomplishments from beginning to end. Completed at the conclusion of the capstone design course(s), the FDR includes everything from problem definition to the creation and validation of a design solution. Given its scope, the FDR often ranges from 100 to 400 pages and includes much of the content documented in the Preliminary Design Report (PDR). To ensure completeness and clarity, a set of guidelines is necessary for preparing the FDR.

19.7.1 Guidelines for the Final Design Report

19.7.1.1 General Format

- **File Format**: Prepare the FDR in PDF format as a single file for consistency and ease of sharing with stakeholders, including the sponsor.
- **Margins**: Use a one-inch margin on all sides of each page.
- **Page Numbers**: Include a page number on every page.
- **Outline**: Follow the prescribed structure outlined below.

19.7.1.2 Title Page

- **Content**: Project title, team number, team name, team logo, team members (and their roles), company sponsor, faculty advisor(s), submission date, and report identification number.
- **Page Number**: Do not insert a page number on the title page.

19.7.1.3 Abstract

- **Content**: A concise summary of the project, objectives, accomplishments, and methods. Should not exceed 500 words and be limited to one page.

19.7.1.4 Table of Contents

- **Content**: List major and minor headings within the report along with page numbers.

19.7.1.5 List of Acronyms

- **Content**: Include all acronyms used in the report.

19.7.1.6 List of Tables

- **Content**: List all tables with their page numbers. Number each table sequentially.

19.7.1.7 List of Figures

- **Content**: List all figures with their page numbers. Number each figure sequentially.

19.7.1.8 Introduction

- **Content**: Project description and problem definition, stated design requirements and expectations, brief chronology of previous work, purpose, scope, and objectives of the design project.

19.7.1.9 Patent Searches

- **Content**: Complete list of all relevant patents searched, with patent drawings where appropriate.

19.7.1.10 Evaluation of the Competition

- **Content**: Market analysis identifying competition and alternatives, and strategy to gain sponsor approval or market advantage.

19.7.1.11 Engineering Design Specifications

- **Content**: Description of how customer requirements and engineering criteria were transcribed into design specifications. Include numerical parameters with units in a tabular format.

19.7.1.12 Conceptual Design

- **Content**: Process for generating design solution concepts. List and illustrate all viable concepts, describe their analysis and evaluation, and detail the Pugh analysis cycles to select top design solutions.

19.7.1.13 Quality Function Deployment (QFD)

- **Content**: QFD analysis to identify important aspects of the project. Include trade-off and competitive analysis, and show the evolution of the QFD throughout the design process.

19.7.1.14 Design for X

- **Content**: Explain how the design was achieved under constraints such as safety, cost, manufacturability, reliability, ergonomics, and environmental impact.

19.7.1.15 Project Specific Details & Analysis

- **Content**: Include data collection activity, engineering analysis, and market analysis. Tailor this section to the specific project type:

- **Product Design Teams**: Market analysis, demand forecasting, cost vs. price information, and surveys of potential users.
- **Process Design Teams**: Flow charts, process diagrams, floor plans, and time studies.
- **National Student Design Competition Teams**: Document work done to fulfill competition requirements beyond typical course requirements.

19.7.1.16 Product Design Details

- **Content**: Development of the chosen concept into the final design. Include drawings, bill-of-materials, process charts, and step-by-step flow descriptions.

19.7.1.17 Engineering Analysis

- **Content**: Detailed explanation of all forms of engineering analysis performed (e.g., process efficiency, thermal, fluids, structural, vibrational, static, dynamic, materials, finite element analysis). Confirm the design's compliance with current knowledge, regulatory requirements, standards, and codes.

19.7.1.18 Build, Manufacture, Create

- **Content**: Steps taken in creating the design solution. For product design, detail the manufacturing analysis and assembly process. For process design, describe the designed process details. Explain scalability to production level.

19.7.1.19 Testing

- **Content**: Verification and validation of the design. Include a detailed test matrix, system-level tests, performance tests, compliance with design specifications, safety, life expectancy, failure modes, disassembly, decommissioning, and disposal. Discuss application of engineering standards (e.g., ASTM, FCC, OSHA).

19.7.1.20 Redesign

- **Content**: Describe redesign cycles based on testing results. Explain rationale and optimization of redesigns, testing procedures, and evolution to the final design. Detail recommendations for future improvements if redesign could not be implemented within time constraints.

19.7.1.21 Project Planning

- **Content**: Project planning and management details. Include a Gantt chart showing timeline, tasks, milestones, team members' roles, and resource allocations. Specify task completion percentages and managerial tools used.

19.7.1.22 Financial Analysis

- **Content**: Detailed description of projected and actual costs, sources of funding, person-hours (and dollar equivalent) spent, cost analysis for scaling up the design solution, market demand, forecast of technological evolution, and return on investment.

19.7.1.23 Operation/Usage

- **Content**: Describe product operation or process usage. Provide an operator's manual or user guide, including safety procedures. Describe maintenance, service requirements, and disposal after the end of its useful life.

19.7.1.24 Additional Considerations

- **Content**: Critical assessment of broader impacts, including:
    - Economic impact
    - Societal impact
    - Ethical considerations
    - Health, ergonomics, and safety considerations
    - Environmental impact and sustainability

19.7.1.25 Conclusions

- **Content**: Document how the design meets specifications, significant findings, potential for commercialization, and next steps.

19.7.1.26 References

- **Content**: List all cited papers, reports, and other references.

19.7.1.27 Appendices

- **Content**: Include supplementary information too bulky for the main report, such as computer programs and detailed drawings. Ensure all items are referenced in the main text.

19.7.2 Professional Formatting Guidelines

19.7.2.1 Footnotes

- Number consecutively using superscript numbers.
- Position flush left at the bottom of the column/page where the first reference appears.
- Use 10 pt. text for footnotes with extra line spacing between text and footnote.

19.7.2.2 Equations

- Display equations separately from the text and centered.
- Number equations consecutively using Arabic numerals in parentheses, positioned flush right along the final baseline of the equation.

19.7.2.3 Graphics

- Include photographs, graphs, and line drawings.
- Number consecutively and caption each graphic.
- Use 11 pt. boldface serif typeface for captions, centered below the graphic.
- Ensure callouts within graphics are no smaller than 9 pt.
- Position graphics close to their references in the text.
- Size graphics for final publication:
    - 7.5 in. across the top of the page
    - 9 x 7.5 in. (L x W) to fit the entire page
    - 6.5 x 8.5 in. (L x W) across the page sideways

19.7.2.4 Tables

- Number consecutively and caption each table.
- Use 11 pt. boldface serif typeface for captions, centered above the table.
- Ensure callouts within tables are no smaller than 9 pt.
- Position tables close to their references in the text.
- Size tables for final publication:
    - 7.5 in. across the top of the page
    - 9 x 7.5 in. (L x W) to fit the entire page
    - 6.5 x 8.5 in. (L x W) across the page sideways

19.7.2.5 Text Citation

- Cite references in numerical order within the text using brackets.
- Example: "It was shown by Prusa [1] that the width of the plume decreases under these conditions."
- For two citations, use a comma [1,2]; for more than two, use a dash [5-7].

19.7.2.6 List of References

- Arrange references in numerical order according to their order of appearance within the text.
- Formats:
    - **Journal Articles**: Last name of each author followed by initials, year of publication, full title in quotes, full name of the publication, volume number (bold), issue number (if any), and inclusive page numbers.

- **Books and Monographs**: Last name of each author followed by initials, year of publication, full title in italics, publisher, city of publication, inclusive page numbers, chapter number (if any).
- **Conference Papers**: Last name of each author followed by initials, year of publication, full title in quotes, individual paper number (if any), full title of publication in italics, editors (if any), publisher, city of publication, volume number (bold), inclusive page numbers.
- **Theses and Technical Reports**: Last name of each author followed by initials, year of publication, full title in quotes, report number (if any), publisher or institution name, city.

19.7.2.7 Sample References

- [1] Ning, X., and Lovell, M. R., 2002, "On the Sliding Friction Characteristics of Unidirectional Continuous FRP Composites," ASME J. Tribol., 124(1), pp. 5-13.
- [2] Barnes, M., 2001, "Stresses in Solenoids," J. Appl. Phys., 48(5), pp. 2000–2008.
- [3] Jones, J., 2000, Contact Mechanics, Cambridge University Press, Cambridge, UK, Chap. 6.
- [4] Lee, Y., Korpela, S. A., and Horne, R. N., 1982, "Structure of Multi-Cellular Natural Convection in a Tall Vertical Annulus," Proc. 7th International Heat Transfer Conference, U. Grigul et al., eds., Hemisphere, Washington, DC, 2, pp. 221–226.
- [5] Hashish, M., 2000, "600 MPa Waterjet Technology Development," High-Pressure Technology, PVP-Vol. 406, pp. 135-140.
- [6] Watson, D. W., 1997, "Thermodynamic Analysis," ASME Paper No. 97-GT-288.
- [7] Tung, C. Y., 1982, "Evaporative Heat Transfer in the Contact Line of a Mixture," Ph.D. thesis, Rensselaer Polytechnic Institute, Troy, NY.
- [8] Kwon, O. K., and Pletcher, R. H., 1981, "Prediction of the Incompressible Flow Over A Rearward-Facing Step," Technical Report No. HTL-26, CFD-4, Iowa State Univ., Ames, IA.
- [9] Smith, R., 2002, "Conformal Lubricated Contact of Cylindrical Surfaces Involved in a Non-Steady Motion," Ph.D. thesis, http://www.cas.phys.unm.edu/rsmith/homepage.html

By following these detailed guidelines, the Final Design Report will be a comprehensive and professional document that effectively communicates the design team's work and accomplishments throughout the entire project.

19.9 Assignments

Assignment 19-1: Create a Poster for your Design Project

Objective

The objective of this assignment is to create a professional and informative poster for your capstone design project. This poster will be displayed at the capstone design showcase to present your project to faculty, industry professionals, and peers. Follow the step-by-step instructions to design, create, and mount your poster on a 36x24 foam board.

Materials Needed
- Provided template for the poster
- 36x24 foam board (to be provided one week before the showcase)
- Access to a computer with design software (e.g., Microsoft PowerPoint, Adobe Illustrator)
- Printer for poster printing
- Glue or adhesive spray for mounting

Step 1: Understanding the Template
1. **Download and Review the Template**:
    - Download the provided template for the poster from the course website or shared drive.
    - Review the template to understand the layout and required sections.
2. **Template Sections**:
    - **Title**: Project title, team members, and advisor(s)
    - **Abstract**: Brief summary of the project, including objectives and significance
    - **Introduction**: Background information and problem statement
    - **Methodology**: Description of the design process and methods used
    - **Results**: Key findings and outcomes of the project
    - **Discussion**: Interpretation of the results and their implications
    - **Conclusion**: Summary of the project and future work
    - **Acknowledgments**: Credits to sponsors, advisors, and contributors

Step 2: Designing the Poster
- **Open the Template**:
    1. Open the provided template in your design software (e.g., Microsoft PowerPoint).
- **Font Sizes and Word Count**:
    1. **Title**: Use a font size of 85-100 points. The title should be easily readable from a distance and limited to around 10 words.

2. **Section Headers**: Use a font size of 48-60 points.
3. **Body Text**: Use a font size of 24-32 points. Each section should contain approximately 150-200 words.
4. **Graphs and Labels**: Ensure labels and text in graphs are at least 18-24 points.
- **Title Section**:
    1. Fill in the project title, team members, and advisor(s) in the title section.
    2. Ensure the title is clear and legible from a distance.
- **Abstract**:
    1. Write a brief abstract summarizing the project's objectives, significance, and key outcomes.
    2. Keep the abstract concise, ideally around 150-200 words.
- **Introduction**:
    1. Provide background information and a clear problem statement.
    2. Explain the context and relevance of the project in 150-200 words.
- **Methodology**:
    1. Describe the design process and methods used in the project.
    2. Include diagrams, flowcharts, or images to illustrate the process. Limit text to 150-200 words.
- **Results**:
    1. Present the key findings and outcomes of the project.
    2. Use charts, graphs, tables, and images to visually represent the data. Limit text to 150-200 words.
- **Discussion**:
    1. Interpret the results and discuss their implications.
    2. Highlight any challenges faced and how they were addressed. Limit text to 150-200 words.
- **Conclusion**:
    1. Summarize the project and suggest future work or improvements.
    2. Emphasize the significance of the findings in 100-150 words.
- **Acknowledgments**:
    1. Credit sponsors, advisors, and any other contributors.
    2. Include logos or names as appropriate. Limit text to 50-100 words.

Step 3: Enhancing the Poster
1. **Visual Design**:
    - Ensure the poster is visually appealing and easy to read.
    - Use high-quality images and graphics.
    - Choose a color scheme that enhances readability and matches the

project's theme.
2. **Guidance on Color Use**:
    - Use contrasting colors for text and background to ensure readability.
    - Limit the use of bright or neon colors that may strain the eyes.
    - Use a consistent color scheme throughout the poster to maintain a cohesive look.
    - Highlight key sections or findings with accent colors.
3. **Fonts and Text**:
    - Use clear, legible fonts (e.g., Arial, Helvetica, Calibri).
    - Avoid overcrowding the poster with too much text.
    - Use bullet points and short paragraphs for clarity.
4. **Consistency**:
    - Maintain a consistent layout and design throughout the poster.
    - Align text and images neatly.

Step 4: Printing and Finishing the Poster
1. **Review and Proofread**:
    - Review the poster for any errors or omissions.
    - Proofread the text for spelling and grammar mistakes.
2. **Print the Poster**:
    - Print the poster using a high-quality printer.
    - Ensure the print dimensions are 36x24 inches.
3. **Mount the Poster**:
    - Obtain the 36x24 foam board provided one week before the showcase.
    - Use glue or adhesive spray to mount the printed poster onto the foam board.
    - Ensure the poster is aligned correctly and smoothly adhered to the board without wrinkles or bubbles.
    - Let the adhesive dry completely before handling the mounted poster.

Step 5: Final Review
1. **Inspect the Mounted Poster**:
    - Inspect the mounted poster for any issues.
    - Make any necessary adjustments to ensure a professional appearance.

Step 6: Submission and Presentation
1. **Submit the Poster**:
    - Submit the mounted poster by the deadline specified by your instructor.
    - Ensure it is ready for display at the capstone design showcase.
2. **Prepare for the Showcase**:

- Be ready to present and discuss your project using the poster at the capstone design showcase.
- Practice explaining your project clearly and concisely to various audiences.

Tips for Success
- Start designing the poster at least three weeks before the design showcase to allow ample time for revisions and printing.
- Collaborate with your team to ensure that all necessary information is included and accurately represented.
- Seek feedback from peers, advisors, and instructors to improve the poster before the final print.
- Ensure all images and graphics are of high resolution to avoid pixelation when printed.

By following these steps and guidelines, you will create a professional and informative poster that effectively communicates your capstone design project. Good luck, and make sure to present your work proudly at the showcase!

Assignment 19-2: Create a Technical Information Sheet for your Design Project

Objective

The objective of this assignment is to create a professional and informative technical information sheet for your capstone design project. This information sheet will be handed out to visitors at the design showcase, providing them with a concise and comprehensive overview of your project. The information sheet must be one double-sided page (letter size) and will not be folded.

Materials Needed
- Computer with design software (e.g., Microsoft Word, Adobe InDesign, Google Docs)
- Printer for printing the information sheets
- High-quality paper for printing (letter size)
- Access to a photocopier or printing service for making copies

Step 1: Initial Planning
1. **Determine Content**:
    - Decide on the key sections and information to include on the information sheet. Suggested sections are:
        - Project Title
        - Team Members
        - Advisor(s)
        - Abstract
        - Introduction
        - Design Process
        - Results and Findings
        - Conclusion and Future Work
        - Contact Information
        - Acknowledgments
2. **Gather Information**:
    - Collect all the necessary information and data for each section. Ensure that all facts and figures are accurate and up-to-date.

Step 2: Designing the Information Sheet
- **Open Design Software**:
    - Open your preferred design software (e.g., Microsoft Word, Adobe InDesign, Google Docs).
- **Set Up the Document**:
    - Set the page size to letter size (8.5 x 11 inches).

19.9 Assignments

- Configure the document to be double-sided.
- **Layout Design**:
  - Divide the page into clear sections using text boxes, columns, or grids.
  - Ensure a clean and professional layout that is easy to read.
- **Title Section**:
  - Place the project title prominently at the top of the front page.
  - Include the names of team members and advisor(s) below the title.
- **Abstract**:
  - Write a brief abstract summarizing the project's objectives, significance, and key outcomes (100-150 words).
- **Introduction**:
  - Provide background information and a clear problem statement (100-150 words).
- **Design Process**:
  - Describe the design process and methodologies used in the project (150-200 words).
  - Include diagrams, flowcharts, or images to illustrate the process.
- **Results and Findings**:
  - Present the key findings and outcomes of the project (150-200 words).
  - Use charts, graphs, tables, and images to visually represent the data.
- **Conclusion and Future Work**:
  - Summarize the project's conclusions and suggest future work or improvements (100-150 words).
- **Contact Information**:
  - Provide contact information for the team (e.g., email addresses, phone numbers).
- **Acknowledgments**:
  - Credit sponsors, advisors, and any other contributors (50-100 words).
- **References** (optional):
  - Include key references and sources if space permits.

Step 3: Enhancing the Information Sheet
- ☐ **Visual Design**:

- Use high-quality images and graphics.
- Choose a color scheme that enhances readability and matches the project's theme.
- **Fonts and Text**:
    - Use clear, legible fonts (e.g., Arial, Helvetica, Calibri).
    - Ensure text size is readable. Use a minimum of 12-point font for body text and larger fonts for headings.
    - Avoid overcrowding the page with too much text.
- **Consistency**:
    - Maintain a consistent layout and design throughout the information sheet.
    - Align text and images neatly.

Step 4: Reviewing and Proofreading
- **Review Content**:
    - Review the information sheet for accuracy and completeness.
    - Ensure all sections are clear and concise.
- **Proofread**:
    - Proofread the text for spelling and grammar mistakes.
    - Ask team members, advisors, or peers to review and provide feedback.

Step 5: Printing and Distribution
- **Finalizing the Document**:
    - Make any necessary revisions based on feedback.
    - Ensure the document is formatted correctly for double-sided printing.
- **Printing the Information Sheets**:
    - Print a test copy to check for any formatting issues.
    - Print approximately 20 copies of the information sheet on high-quality paper.
- **Distributing the Information Sheets**:
    - Ensure each team member has access to the printed information sheets.
    - Hand out the information sheets to visitors during the design showcase.

Step 6: Submission and Preparation
- **Submit the Information Sheet**:
    - Submit a digital copy of the information sheet to your instructor at

least three weeks before the design showcase.
- □ **Prepare for the Showcase**:
    - o Be ready to present and discuss your project using the information sheet at the capstone design showcase.
    - o Practice explaining your project clearly and concisely to various audiences.

By following these steps, you will create a professional and informative technical information sheet that effectively communicates your capstone design project. Ensure that your information sheet is visually appealing, easy to read, and accurately represents your project's key elements. Good luck, and make sure to present your work proudly at the showcase!

Bibliography

A. Buzzetto-More, N., Julius Alade, A., 2006. Best Practices in e-Assessment. JITE:Research 5, 251–269. https://doi.org/10.28945/246

Adams, D.F., Odom, E.M., 1987. Testing of single fibre bundles of carbon/carbon composite materials. Composites 18, 381–385. https://doi.org/10.1016/0010-4361(87)90362-4

Al-Thani, S.B.J., Abdelmoneim, A., Cherif, A., Moukarzel, D., Daoud, K., 2016. Assessing general education learning outcomes at Qatar University. Jnl of Applied Research in HE 8, 159–176. https://doi.org/10.1108/JARHE-03-2015-0016

Applied Imagination - Wikipedia [WWW Document], n.d. URL https://en.wikipedia.org/wiki/Applied_Imagination (accessed 7.12.20).

Atadero, R.A., Rambo-Hernandez, K.E., Balgopal, M.M., 2015. Using Social Cognitive Career Theory to Assess Student Outcomes of Group Design Projects in Statics: SCCT and Student Outcomes of Statics Projects. J. Eng. Educ. 104, 55–73. https://doi.org/10.1002/jee.20063

Benavides, E., 2011. Advanced Engineering Design. Woodhead Publishing.

Buchsbaum, A., Rey, C., n.d. Jean-Yves Beziau Federal University of Rio de Janeiro Rio de Janeiro, RJ Brazil 623.

Caple, M., Wild, J., Maslen, E., Nagel, J., 2015. Design of a controller for a hybrid bearing system, in: 2015 Systems and Information Engineering Design Symposium. Presented at the 2015 Systems and Information Engineering Design Symposium, IEEE, Charlottesville, VA, USA, pp. 236–239. https://doi.org/10.1109/SIEDS.2015.7116980

Carnevalli, J.A., Miguel, P.A.C., Calarge, F.A., 2010. Axiomatic design application for minimising the difficulties of QFD usage. International Journal of Production Economics 125, 1–12. https://doi.org/10.1016/j.ijpe.2010.01.002

CATIA [WWW Document], n.d. URL https://www.3ds.com/products-services/catia/?wockw=card_content_cta_1_url%3A%22https%3A%2F%2Fblogs.3ds.com%2Fcatia%2F%22

Cordon, D., Clarke, E., Westra, L., Allen, N., Cunnington, M., Drew, B., Gerbus, D., Klein, M., Walker, M., Odom, E.M., Rink, K.K., Beyerlein, S.W., 2002. Shop orientation to enhance design for manufacturing in capstone projects, in: 32nd Annual Frontiers in Education. Presented at the Conference on Frontiers in Education, IEEE, Boston, MA, USA, pp. F4D-6-F4D-11. https://doi.org/10.1109/FIE.2002.1158229

Council on Higher Education (South Africa), 2011. Work-integrated learning: good practice guide. Council on Higher Education, Pretoria, South Africa.

Creativity Unbound [WWW Document], n.d. . FourSight. URL https://foursightonline.com/product/creativity-unbound/ (accessed 8.16.20).

Criteria for Accrediting Engineering Programs, 2019 – 2020 | ABET [WWW Document], n.d. URL https://www.abet.org/accreditation/accreditation-criteria/criteria-for-accrediting-engineering-programs-2019-2020/ (accessed 6.10.20).

Davis, D.C., Gentili, K.L., Trevisan, M.S., Calkins, D.E., 2002. Engineering Design Assessment Processes and Scoring Scales for Program Improvement and Accountability. Journal of Engineering Education 91, 211–221. https://doi.org/10.1002/j.2168-9830.2002.tb00694.x

Deveci, T., Nunn, R., 2018. COMM151: A PROJECT-BASED COURSE TO ENHANCE ENGINEERING STUDENTS' COMMUNICATION SKILLS. ESPEAP 6, 027. https://doi.org/10.22190/JTESAP1801027D

Diefes-Dux, H.A., Moore, T., Zawojewski, J., Imbrie, P.K., Follman, D., 2004. A framework for posing open-ended engineering problems: model-eliciting activities, in: 34th Annual Frontiers in Education, 2004. FIE 2004. Presented at the 34th Annual Frontiers in Education, 2004. FIE 2004., IEEE, Savannah, GA, USA, pp. 455–460. https://doi.org/10.1109/FIE.2004.1408556

Dollar, A.M., Kerdok, A.E., Diamond, S.G., Novotny, P.M., Howe, R.D., 2005. Starting on the Right Track: Introducing Students to Mechanical Engineering With a Project-Based Machine Design Course, in: Innovations in Engineering Education: Mechanical Engineering Education, Mechanical Engineering/Mechanical Engineering Technology Department Heads. Presented at the ASME 2005 International Mechanical Engineering Congress and Exposition, ASMEDC, Orlando, Florida, USA, pp. 363–371. https://doi.org/10.1115/IMECE2005-81929

Duarte, B.B., de Castro Leal, A.L., de Almeida Falbo, R., Guizzardi, G., Guizzardi, R.S.S., Souza, V.E.S., 2018. Ontological foundations for software requirements with a focus on requirements at runtime. AO 13, 73–105. https://doi.org/10.3233/AO-180197

Dynn, C.L., Agogino, A.M., Eris, O., Frey, D.D., Leifer, L.J., 2006. Engineering design thinking, teaching, and learning. IEEE Eng. Manag. Rev. 34, 65–65. https://doi.org/10.1109/EMR.2006.1679078

Eberle, B., 2008. Scamper: Creative Games and Activities for Imagination Development. Prufrock Press.

Egelhoff, C., Odom, E., 2001. Advanced learning made as easy as ABC: an example using design for fatigue of machine elements subjected to simple and combined loads, in: 31st Annual Frontiers in Education Conference. Impact on Engineering and Science Education. Conference Proceedings (Cat. No.01CH37193). Presented at the 31st Annual Frontiers in Education Conference. Impact on Engineering and Science Education., IEEE, Reno, NV, USA, pp. T4B-7-T4B-12. https://doi.org/10.1109/FIE.2001.963934

Egelhoff, C.F., Odom, E.M., 1999. Machine design: where the action should be, in: FIE'99 Frontiers in Education. 29th Annual Frontiers in Education Conference. Designing the Future of Science and Engineering Education. Conference Proceedings (IEEE Cat. No.99CH37011. Presented at the IEEE Computer Society Conference on Frontiers in Education, Stripes Publishing L.L.C, San Juan, Puerto Rico, p. 12C5/13-12C5/18. https://doi.org/10.1109/FIE.1999.841644

Egelhoff, C.J., Odom, E.M., Wiest, B.J., 2010. Application of modern engineering tools in the analysis of the stepped shaft: Teaching a structured problem-solving approach using energy techniques, in: 2010 IEEE Frontiers in Education Conference (FIE). Presented at the 2010 IEEE Frontiers in Education Conference (FIE), IEEE, Arlington, VA, USA, pp. T1C-1-T1C-6. https://doi.org/10.1109/FIE.2010.5673504

Facebook, Twitter, options, S. more sharing, Facebook, Twitter, LinkedIn, Email, URLCopied!, C.L., Print, 1999. Mars Probe Lost Due to Simple Math Error [WWW Document]. Los Angeles Times. URL https://www.latimes.com/archives/la-xpm-1999-oct-01-mn-17288-story.html (accessed 8.16.20).

Farid, A.M., Suh, N.P. (Eds.), 2016. Axiomatic Design in Large Systems. Springer International Publishing, Cham. https://doi.org/10.1007/978-3-319-32388-6

Bibliography

Griffin, P.M., Griffin, S.O., Llewellyn, D.C., 2004. The Impact of Group Size and Project Duration on Capstone Design. Journal of Engineering Education 93, 185–193. https://doi.org/10.1002/j.2168-9830.2004.tb00805.x

Howe, S., Rosenbauer, L., Poulos, S., n.d. 2015 Capstone Design Survey – Initial Results 4.

Ibrahim, A., Kurfess, T.R., Agogino, A., Alani, F., Burgess, S., Chen, W.-F., Lantada, A.D., Ertugrul, N., Felder, R., Genalo, L., Gillet, D., Hernandez, W., n.d. FOUNDER AND FIRST EDITOR-IN-CHIEF MICHAEL S. WALD (1932–2008) 1.

Iowa State University, Carter, R., Strader, T., Drake University, Rozycki, J., Drake University, Root, T., Drake University, 2015. Cost Structures of Information Technology Products and Digital Products and Services Firms: Implications for Financial Analysis. JMWAIS 2015, 5–19. https://doi.org/10.17705/3jmwa.00002

Leaman, E.J., Cochran, J.R., Nagel, J.K., 2014. Design of a two-phase solar and fluid-based renewable energy system for residential use, in: 2014 Systems and Information Engineering Design Symposium (SIEDS). Presented at the 2014 Systems and Information Engineering Design Symposium (SIEDS), IEEE, Charlottesville, VA, USA, pp. 193–197. https://doi.org/10.1109/SIEDS.2014.6829917

Morell, L., 2015. Disrupting engineering education to better address societal needs, in: 2015 International Conference on Interactive Collaborative Learning (ICL). Presented at the 2015 International Conference on Interactive Collaborative Learning (ICL), IEEE, Firenze, Italy, pp. 1093–1097. https://doi.org/10.1109/ICL.2015.7318184

Nagel, J., 2013. Guard Cell and Tropomyosin Inspired Chemical Sensor. Micromachines 4, 378–401. https://doi.org/10.3390/mi4040378

Nagel, J.K.S., Nagel, R.L., Eggermont, M., 2013. Teaching Biomimicry With an Engineering-to-Biology Thesaurus, in: Volume 1: 15th International Conference on Advanced Vehicle Technologies; 10th International Conference on Design Education; 7th International Conference on Micro- and Nanosystems. Presented at the ASME 2013 International Design Engineering Technical Conferences and Computers and Information in Engineering Conference, American Society of Mechanical Engineers, Portland, Oregon, USA, p. V001T04A017. https://doi.org/10.1115/DETC2013-12068

Nagel, J.K.S., Nagel, R.L., Stone, R.B., 2011. Abstracting biology for engineering design. IJDE 4, 23. https://doi.org/10.1504/IJDE.2011.041407

Nagel, Jacquelyn K.S., Nagel, R.L., Stone, R.B., McAdams, D.A., 2010. Function-based, biologically inspired concept generation. AIEDAM 24, 521–535. https://doi.org/10.1017/S0890060410000375

Nagel, J.K.S., Stone, R.B., 2012. A computational approach to biologically inspired design. AIEDAM 26, 161–176. https://doi.org/10.1017/S0890060412000054

Nagel, Jacquelyn K. S., Stone, R.B., McAdams, D.A., 2010. An Engineering-to-Biology Thesaurus for Engineering Design, in: Volume 5: 22nd International Conference on Design Theory and Methodology; Special Conference on Mechanical Vibration and Noise. Presented at the ASME 2010 International Design Engineering Technical Conferences and Computers and Information in Engineering Conference, ASMEDC, Montreal, Quebec, Canada, pp. 117–128. https://doi.org/10.1115/DETC2010-28233

Odom, E.M., Adams, D.F., 1990. Failure modes of unidirectional carbon/epoxy composite compression specimens. Composites 21, 289–296. https://doi.org/10.1016/0010-4361(90)90343-U

Odom, E.M., Beyerlein, S.W., Tew, B.W., Smelser, R.E., Blackketter, D.M., 1999. Idaho Engineering Works: a model for leadership development in design education, in: FIE'99 Frontiers in Education. 29th Annual Frontiers in Education Conference. Designing the Future of Science and Engineering Education. Conference Proceedings (IEEE Cat. No.99CH37011. Presented at the IEEE Computer Society Conference on Frontiers in Education, Stripes Publishing L.L.C, San Juan, Puerto Rico, p. 11B2/21-11B2/24. https://doi.org/10.1109/FIE.1999.839217

Odom, E.M., Egelhoff, C.J., 2011. Teaching deflection of stepped shafts: Castigliano's theorem, dummy loads, heaviside step functions and numerical integration, in: 2011 Frontiers in Education Conference (FIE). Presented at the 2011 Frontiers in Education Conference (FIE), IEEE, Rapid City, SD, USA, pp. F3H-1-F3H-6. https://doi.org/10.1109/FIE.2011.6143039

Online, S.B., n.d. Advanced Engineering Design [WWW Document]. URL https://learning.oreilly.com/library/view/advanced-engineering-design/9780857090935/ (accessed 6.6.20).

ORACLE PRIMAVERA P6 SOFTWARE [WWW Document], n.d. URL https://globalpm.com/products/oracle-primavera-p6-software/

Osborne, A.F., 1953. Applied imagination; principles and procedures of creative thinking. (Book, 1953) [WorldCat.org] [WWW Document]. URL https://www.worldcat.org/title/applied-imagination-principles-and-procedures-of-creative-thinking/oclc/641122686 (accessed 7.12.20).

Pahl, A.K., Newnes, L., McMahon, C., 2007. A generic model for creativity and innovation: overview for early phases of engineering design. JDR 6, 5. https://doi.org/10.1504/JDR.2007.015561

Pembridge, J.J., Paretti, M.C., 2019. Characterizing capstone design teaching: A functional taxonomy. J. Eng. Educ. 108, 197–219. https://doi.org/10.1002/jee.20259

Ribeiro, A.L., Bittencourt, R.A., 2018. A PBL-Based, Integrated Learning Experience of Object-Oriented Programming, Data Structures and Software Design, in: 2018 IEEE Frontiers in Education Conference (FIE). Presented at the 2018 IEEE Frontiers in Education Conference (FIE), IEEE, San Jose, CA, USA, pp. 1–9. https://doi.org/10.1109/FIE.2018.8659261

Ritter, S.M., Mostert, N., 2017. Enhancement of Creative Thinking Skills Using a Cognitive-Based Creativity Training. J Cogn Enhanc 1, 243–253. https://doi.org/10.1007/s41465-016-0002-3

Romaniuk, R.S., 2012. Communications, Multimedia, Ontology, Photonics and Internet Engineering 2012. International Journal of Electronics and Telecommunications 58, 463–478. https://doi.org/10.2478/v10177-012-0061-z

S. Nagel, J.K., W., F., 2012. Hybrid Manufacturing System Design and Development, in: Abdul Aziz, F. (Ed.), Manufacturing System. InTech. https://doi.org/10.5772/35597

Salustri, F.A., Eng, N.L., Weerasinghe, J.S., 2008. Visualizing Information in the Early Stages of Engineering Design. Computer-Aided Design and Applications 5, 697–714. https://doi.org/10.3722/cadaps.2008.697-714

Stroble, Jacquelyn K., Stone, R.B., Watkins, S.E., 2009. Assessing How Digital Design Tools Affect Learning of Engineering Design Concepts, in: Volume 8: 14th Design for Manufacturing and the Life Cycle Conference; 6th Symposium on International Design and Design Education; 21st International Conference on Design Theory and

Methodology, Parts A and B. Presented at the ASME 2009 International Design Engineering Technical Conferences and Computers and Information in Engineering Conference, ASMEDC, San Diego, California, USA, pp. 467–476. https://doi.org/10.1115/DETC2009-86708

Stroble, J.K., Stone, R.B., Watkins, S.E., 2009. An overview of biomimetic sensor technology. Sensor Review 29, 112–119. https://doi.org/10.1108/02602280910936219

Swartwout, M., Kitts, C., Twiggs, R., Kenny, T., Ray Smith, B., Lu, R., Stattenfield, K., Pranajaya, F., 2008. Mission results for Sapphire, a student-built satellite. Acta Astronautica 62, 521–538. https://doi.org/10.1016/j.actaastro.2008.01.009

The Gantt chart, a working tool of management : Clark, Wallace, 1880-1948 : Free Download, Borrow, and Streaming [WWW Document], n.d. . Internet Archive. URL https://archive.org/details/cu31924004570853 (accessed 7.6.20).

The_effectiveness_of_cooperative_problem.pdf, n.d.

Todd, R.H., Magleby, S.P., Sorensen, C.D., Swan, B.R., Anthony, D.K., 1995. A Survey of Capstone Engineering Courses in North America. Journal of Engineering Education 84, 165–174. https://doi.org/10.1002/j.2168-9830.1995.tb00163.x

Truong, H.T.X., Odom, E.M., Egelhoff, C.J., Burns, K.L., 2011. Using Modern Engineering Tools to Efficiently Solve Challenging Engineering Design Problems: Analysis of the Stepped Shaft 10.

Tuckman, B.W., Jensen, M.A.C., 1977. Stages of Small-Group Development Revisited. Group & Organization Studies 2, 419–427. https://doi.org/10.1177/105960117700200404

Wenhao Huang, D., Diefes-Dux, H., Imbrie, P.K., Daku, B., Kallimani, J.G., 2004. Learning motivation evaluation for a computer-based instructional tutorial using ARCS model of motivational design, in: 34th Annual Frontiers in Education, 2004. FIE 2004. Presented at the 34th Annual Frontiers in Education, 2004. FIE 2004., IEEE, Savannah, GA, USA, pp. 65–71. https://doi.org/10.1109/FIE.2004.1408466

Why ABET Accreditation Matters | ABET [WWW Document], n.d. URL https://www.abet.org/accreditation/what-is-accreditation/why-abet-accreditation-matters/ (accessed 6.10.20).

Williams, C.B., Gero, J., Lee, Y., Paretti, M., 2010. Exploring Spatial Reasoning Ability and Design Cognition in Undergraduate Engineering Students, in: Volume 6: 15th Design for Manufacturing and the Lifecycle Conference; 7th Symposium on International Design and Design Education. Presented at the ASME 2010 International Design Engineering Technical Conferences and Computers and Information in Engineering Conference, ASMEDC, Montreal, Quebec, Canada, pp. 669–676. https://doi.org/10.1115/DETC2010-28925

Xiao, A., Park, S.S., Freiheit, T., 2011. A COMPARISON OF CONCEPT SELECTION IN CONCEPT SCORING AND AXIOMATIC DESIGN METHODS. PCEEA. https://doi.org/10.24908/pceea.v0i0.3769

zotero / Insert Items from Amazon [WWW Document], n.d. URL http://zotero.pbworks.com/w/page/5511972/Insert%20Items%20from%20Amazon (accessed 8.16.20).

Index

A

accuracy, 69, 73, 80, 82, 114, 142, 249, 261, 292
across, 56, 116, 120–121, 135, 140, 164, 177, 196–197, 201, 223–224, 283
adaptive design, 198
additional costs, 134, 147
aesthetics, 54, 140, 142, 175
aesthetics,, 140, 142
aircraft, 138, 167
algorithms, 102–103, 107–108, 135, 139, 153
alloys, 45
alternative solutions, 246
aluminum, 58, 62, 154–155, 219, 239
American Society of Mechanical, 297
analysis, 31, 37, 59–60, 62, 64–65, 68, 97, 119, 122–123, 133, 135–136, 143, 145–146, 148, 154, 156, 172, 194, 240, 243, 253, 255, 257, 262, 266–269, 280–282, 296
analysis, 31, 37, 59–60, 62, 64–65, 68, 97, 119, 122–123, 133, 135–136, 143, 145–146, 148, 154, 156, 172, 194, 240, 243, 253, 255, 257, 262, 266–269, 280–282, 296
approach, 30, 33, 39–43, 57, 61–62, 64, 68–69, 90–91, 93–94, 96, 98, 100, 103, 110, 118, 125–127, 132, 143, 153, 155, 163–164, 166, 169, 179, 184, 187, 189, 195, 199, 202, 231, 253, 269, 296–297
architecture, 182, 201
assembly, 31–32, 34, 38, 45, 54, 56–57, 59–64, 67–69, 75, 78, 80, 82, 87–89, 91, 97, 126–127, 129, 144, 156, 235, 240, 263–265, 268, 281
assembly processes, 56, 68
automated, 78, 97, 101, 110, 118, 143, 267
availability, 33, 37–39, 57, 59, 62, 88, 131, 142, 197

B

based on, 29, 31, 33–35, 39, 41, 46, 49, 59, 73, 76–77, 84, 87–88, 96, 100, 122–123, 125, 128–129, 134, 136, 142, 145, 151–153, 155, 157, 159–160, 173, 177, 182, 203, 213, 215, 218, 229, 244, 247, 250, 254, 281, 292
bearings, 47
benefits, 104, 141, 146, 148, 187, 248, 274
benefits of, 148, 248
best practices, 69, 75, 86, 95–96, 98, 101, 103–104, 110–111, 117–118, 164, 208, 213–214, 219, 223, 229, 253, 269
budgeting, 82
building, 36, 40, 43, 59, 75–76, 78, 83, 145, 170, 181, 183, 188, 248

C

calculating, 267
carbon footprint, 43, 182, 221
cast iron, 45
casting, 56, 59
categories, 203, 238
Center, 81
ceramics, 43
characteristics, 61, 74, 111, 140, 154, 170
characteristics as, 170

checking, 247
Code, 228
codes, 168, 190, 195, 203, 228, 281
codes of ethics, 190, 228
Commission, 199
competition, 257, 280–281
complexity, 40, 61–62, 68, 77, 91, 124, 140, 254
complexity of, 140, 254
components of, 93
composite materials, 139, 295
compromise, 62, 244
concept generation, 297
configurations, 99, 154
conflict, 130, 234, 252
conflicts of interest, 190
conformance, 140–143
connections, 68, 176, 188–189
considerations, 41, 43, 49, 57, 60, 63, 163–164, 167–175, 177, 179, 184–185, 188–190, 192–198, 203–204, 206, 208–209, 216, 221, 223, 228, 230, 233, 235–238, 241–242, 244, 258, 282
considering, 43, 54, 56, 59, 62, 91, 124, 132–133, 140, 163, 168, 172, 178, 187, 200
constraints, 37–40, 42, 46, 49, 60, 92, 124, 128–130, 132–133, 136–140, 205, 210, 215, 220, 225, 230, 235, 240, 243, 280–281
Control charts, 143
controlling, 174
of controls, 167, 170, 209
cooling, 56, 75
copyright, 249
copyright, 249
corrosion, 46–48
corrosion resistance, 46, 48
cost estimates, 257

Cost evaluation, 144–145, 148
cost evaluation, 144–145, 148
costs, 33–34, 40, 44, 54, 56, 58–59, 61–62, 64, 69–70, 79, 83–84, 88, 95, 134–135, 145–148, 152, 200, 238, 243–246, 252, 257, 282
costs and, 54, 61–62, 69, 146, 246
costs of, 146
creating, 29, 38, 55, 61, 71, 77, 82, 93–94, 163, 169, 172, 175, 179, 186, 189–190, 195, 198, 200–202, 281
creating, 29, 38, 55, 61, 71, 77, 82, 93–94, 163, 169, 172, 175, 179, 186, 189–190, 195, 198, 200–202, 281
creative thinking, 298
Creativity, 92, 295, 298
critical thinking, 195, 253, 272
curves, 171
customer requirements and, 280
customer satisfaction, 142, 144
cycle, 127, 147, 158, 164, 179, 199, 218, 221–222, 238, 240

D

decisions, 46, 93, 98, 103, 110, 125, 129–130, 133, 142, 145, 148, 152, 164, 171, 177, 180, 187, 190, 196–197, 199, 229, 234, 238, 249, 254–255, 259, 263
defects, 44, 56, 64, 84, 140, 143–144
defensive, 250
defined, 30–32, 87, 102, 108–109, 199
design, 29–33, 35–43, 45, 49–50, 54–66, 68–69, 72, 75–80, 82–83, 86–89, 91–104, 106, 110–111, 113–114, 117–120, 122–145, 147–148, 152–160, 163–192, 194–216, 218–221, 223–226, 228–231, 233–236, 238–241, 243–251, 253–259, 261–264,

266–270, 273–276, 279–282, 284, 286–293, 295–299

design for, 57, 176, 184, 188, 210, 219, 295–296

design with, 179, 187

in design, 50, 97, 101, 110, 136, 145, 171, 177, 181, 186, 192, 199–200, 298

design, 29–33, 35–43, 45, 49–50, 54–66, 68–69, 72, 75–80, 82–83, 86–89, 91–104, 106, 110–111, 113–114, 117–120, 122–145, 147–148, 152–160, 163–192, 194–216, 218–221, 223–226, 228–231, 233–236, 238–241, 243–251, 253–259, 261–264, 266–270, 273–276, 279–282, 284, 286–293, 295–299

design as, 179, 243–244, 247–249, 251

Design of Experiments, 144

design parameters, 69, 119–120, 136, 140, 144, 153–155

design process, 29, 38, 43, 56, 61, 75, 78, 80, 93–96, 98, 101, 125, 129, 136, 142, 144, 148, 153, 156, 163–164, 169, 171–172, 174, 179, 185–186, 189–190, 192, 195, 198, 206–207, 210–211, 216, 221, 226, 231, 236, 241, 243–246, 254–255, 259, 264, 269, 274, 280, 286–287, 291

design review, 63

design variables, 136–138

development, 29–32, 35–37, 39, 41–42, 61, 64, 92–94, 96–98, 100–101, 103–104, 106, 111, 114, 118, 140, 143, 173, 181, 188, 194–196, 199, 201, 243, 245–249, 252, 254, 260, 263–264, 298

development of, 29, 31–32, 39, 41, 94, 101, 106, 114, 118, 195–196, 201, 245–246

development costs, 246

development steps, 254

dimensions., 138, 155

distribution, 173, 186–187, 196–197, 209

distribution), 186

Documentation, 31, 35, 38, 68, 79, 83, 85, 89, 92, 96, 98, 100–101, 103–105, 107, 110, 113, 117, 128, 131, 133, 146, 160, 203, 207–208, 213, 218, 221, 223, 226, 228, 231, 233, 236, 238, 241, 253–256, 258, 260, 262, 264, 266–268, 270, 272, 274, 276, 278, 280, 282, 284, 286, 288, 290, 292, 294

documentation, 31–32, 35, 38–39, 68, 78–79, 82, 87, 89, 96, 98–99, 101, 106, 110–111, 117, 129, 131, 135, 145–146, 189, 206, 211, 216, 221, 224, 226, 235–236, 253–255, 258–259, 263–264, 266, 269

documenting, 34, 106, 109, 118, 120, 124, 204, 206, 263–264, 266

durability, 38, 46, 48, 54, 57, 73, 112, 114, 125, 127, 139–142, 180, 200, 239, 244, 251

E

ease of, 41, 47, 54, 70, 109, 112, 115, 117, 168, 180, 189, 255, 279

ease of fabrication, 47

ease of inspection, 168

ease of use, 70, 109, 112, 117

economy, 164, 180, 184, 191, 200

electronic devices, 180

elements of, 30, 170

elements of, 30, 170

email, 256, 291

embodied energy, 181–182

energy use, 181, 219, 239
engineering societies, 190
Engineers, 97, 102, 163–166, 168, 170, 174–175, 179–183, 185–189, 191–194, 196–198, 229, 297
environmental, 38–39, 45, 49, 107, 109, 112, 114, 121, 128, 163–164, 168–170, 177, 179–180, 182, 184–185, 187, 191, 194–197, 199, 201–202, 204, 218–222, 230, 234, 236, 238, 240, 243–244, 251, 257, 280
Environmental impact, 196, 282
environmental impact, 109, 128, 179–180, 184, 194, 219, 230, 240, 243–244, 280
Environmental Protection, 219
environments, 46, 48, 93, 103, 107–108, 112, 116, 135, 171–177, 179, 185, 189
ergonomics, 163, 169–174, 208–209, 211, 280, 282
erosion, 44
estimating, 59, 146
estimating, 59, 146
ethical, 163–164, 169, 188, 190–195, 197, 202, 224, 226, 228–232, 244, 248–252, 257
ethics, 190, 194–195, 228, 243
Evaluation, 31, 76–77, 92, 94, 104–105, 107, 122, 144–147, 160, 165, 172, 177, 207, 212, 217, 222, 227, 232, 237, 242, 280
evaluation, 41, 77, 104, 113, 144–145, 148, 152, 172–173, 179, 280, 299
evaluation., 104, 148, 179
evolution, 128, 137, 254, 263, 280–282
example, 57–58, 61, 66–67, 80, 88, 97, 101, 106, 114, 127, 140, 148–149, 154, 166, 200, 296
examples, 65, 92

Excel, 85, 121
expenses, 34–35, 56, 132, 144, 146–148, 244
experience, 76–79, 86, 91, 106, 131, 159, 172, 176, 210, 269
extrusion, 71, 74

F

factors, 54, 57–58, 61, 91, 129, 132, 140, 144, 153–154, 163–164, 169–171, 177–178, 191, 195–197, 200, 233, 236–237, 257
failure modes, 94, 119, 143–144, 281
features, 30, 57, 68, 70–72, 74, 91–92, 124, 127, 129, 131, 140–142, 153, 168, 174, 176–178, 204–207, 209–211, 213–216, 218–227, 229–232, 234–243, 245–246, 263–264, 274–275
finite element analysis, 281
fishbone diagrams, 65, 68
fit, 57, 68, 88, 91, 167, 169, 172, 264, 283
fixes, 123
flowcharts, 287, 291
fracture toughness, 47
from, 29–33, 36–37, 39, 41, 55, 61–62, 64, 68–70, 74, 77–78, 80, 83–84, 86–87, 91–92, 96, 100, 109–110, 113, 115, 117–118, 122–124, 127, 129, 131, 133–134, 136, 140–141, 146–147, 153, 155–156, 158–159, 165, 167–169, 174–175, 179, 181, 184, 195, 198–199, 201, 204–206, 211, 213–214, 216, 218, 220–221, 225–226, 230–231, 235–236, 238–241, 246–247, 249–253, 256, 259–262, 264, 269, 272–273, 279, 283, 286–287, 289, 299

functional requirements, 77, 111, 154

G

Gantt charts, 256
Garvin's eight dimensions of, 144
genetic algorithms, 139
guidelines, 32, 173, 195, 197, 208, 211, 215, 219, 224, 247, 253–255, 259, 261–262, 274, 276, 279, 284, 289

H

handling, 49, 58, 140, 204, 208–209, 215, 245, 288
hardness, 46–47, 56
hazardous materials, 166, 168, 213, 218
HDPE, 45
hearing, 175, 213–215
human factors engineering, 169

I

ideas, 30, 91, 254, 272–273
of ideas, 254, 272
identification, 39, 93, 165, 279
identifying customer needs, 142
IEEE, 228, 295–299
imagination, 194, 263, 298
impact resistance, 46–48
importance, 32, 54, 76, 82, 132, 144, 169, 189–190, 195, 206, 211, 216, 221, 226, 236, 241
individuals, 37, 84–85, 208, 224, 229, 248
inequality constraints, 137
Information, 80, 171, 175, 258, 269, 273–277, 290–292, 295, 297–299
information on, 49, 95, 104
injection molding, 58–59

innovation and, 180
innovative, 64, 75, 92, 140, 164, 185, 190–191, 246–247, 272
inspection, 56, 59–60, 143, 168
integral, 94, 98, 129, 148, 163, 194, 198, 255, 259
intellectual property, 193
intellectual property rights, 193
interfaces, 36, 68, 165, 167, 169, 171, 173, 176, 186, 189, 209, 213, 215, 224, 226
Internet, 169, 192, 200, 298–299
introduction, 259
Ishikawa diagrams, 65
issues, 32–35, 39–41, 59–60, 62–63, 66, 73–75, 78, 84–85, 88, 92–96, 98, 100–101, 103–107, 109–113, 117–118, 121–125, 129–131, 133, 135, 147, 156–158, 168, 175, 185–186, 192–194, 205, 220, 230, 262, 264, 288, 292
iteration, 30, 41, 62, 75, 79, 124–125, 158–160

J

judgment, 193

K

knowledge, 75–77, 79, 97, 101, 106, 123, 164, 172, 174, 186, 189, 195–196, 198, 246, 253, 255, 259, 269, 281

L

labor costs, 54, 58–59
lead times, 54, 80, 88, 143
leadership, 196, 251, 298

learning curves, 171
libraries, 36
life cycle, 147, 164, 179, 199, 218, 221–222, 238, 240
limit, 126, 130, 261
loyalty, 142, 193
lubrication, 97–98

M

machinability, 46, 57
machining, 31, 46, 55, 57–60, 62–64
maintenance, 44, 54, 56, 61, 98, 103, 115, 141, 145, 147, 168, 179, 184, 189, 200–201, 243, 263–264, 282
maintenance and, 145, 147, 189
manuals, 32, 142, 167, 257
manufacturability, 54, 56–57, 61–64, 91, 257, 280
manufacturing, 32, 44, 46, 54–64, 68–71, 79, 82, 114, 141, 143–144, 147–148, 151, 154, 167–168, 174, 180–182, 197, 210, 220–221, 235, 239–241, 244, 247, 251, 257, 281, 295
manufacturing costs, 79, 257
manufacturing process, 55–61, 69, 82, 143, 151, 251
manufacturing processes, 32, 44, 54, 58–59, 63, 69, 180–182, 247
marketing, 142
markets, 188, 200, 233
material properties, 45, 49, 129, 136, 138–139
for materials, 34, 88
materials, 30, 32–34, 37–40, 42–50, 54–60, 69–71, 73, 79, 82–83, 86–89, 93, 95, 98–99, 113, 115–116, 126–132, 134, 138–139, 142–143, 145–147, 154–155, 157, 166, 168–169, 171, 174, 180–184, 199, 201, 203–205, 213–215, 218–219, 221, 224, 234, 238–241, 246–247, 252, 255–257, 264, 281, 295
of materials, 43–44, 49–50, 56–57, 59, 70–71, 79, 83, 86, 89, 138, 143, 154, 174, 224, 239
mean time between failures, 141
measurement, 126, 143
Mechanical Engineering, 296
meetings, 34, 133, 186, 254, 273
meetings, 34, 133, 186, 254, 273
method, 55, 58–59, 64, 69, 92, 105, 126, 137, 144, 147, 153, 218, 255
Methods, 55, 94, 104–105, 127, 137, 146, 153, 172, 186
methods, 54–55, 62, 93, 105, 110, 113, 116, 119, 129, 137, 139, 143–144, 153, 155–157, 165, 172–173, 186, 206, 256, 269, 279, 286–287
minimizing, 64, 79, 107, 109–110, 136, 138, 166, 174, 179, 230
mistakes, 64–65, 68, 78, 91–92, 288, 292
modeling, 33, 35, 37, 59, 139, 153, 172, 184, 257
models, 35, 38, 58, 69, 75, 80, 94, 139–140, 146, 148, 153, 157, 159, 180, 200, 224, 240, 257, 265, 267
modular, 68, 78, 184, 200, 224, 247
modules, 56, 61
multi-objective optimization, 139
multiple, 31, 35, 61, 63, 71, 84, 108, 114, 119–121, 127, 139, 152, 157, 165, 196

N

nonlinear, 137
nuclear power, 167

O

Office, 38
optimization, 30, 76, 78, 136–140, 152–155, 171, 267, 281
optimization tools, 140
organizational structures, 171
organizations, 172, 186, 198, 226
original design, 89, 244
outsourcing, 58, 63, 78
overview, 273–274, 290, 298–299
overview, 273–274, 290, 298–299
of ownership, 44, 141, 147

P

parameters, 69, 74, 96, 100, 119–120, 122, 125, 136, 139–140, 144, 151, 153–157, 166, 267, 280
partnerships, 58, 196
Parts, 61, 63, 79, 299
of parts, 54, 61
parts, 38, 47–48, 54–57, 61–63, 67, 71–72, 75, 79, 82–83, 88, 94, 126–127, 130–131, 142–143, 160, 165, 204, 207, 232, 263–264, 268, 273
parts for, 75
parts, 38, 47–48, 54–57, 61–63, 67, 71–72, 75, 79, 82–83, 88, 94, 126–127, 130–131, 142–143, 160, 165, 204, 207, 232, 263–264, 268, 273
patents, 70, 256, 280
perceived quality, 140, 142
performance characteristics, 61, 111
performance metrics, 30, 87, 108, 112, 140, 274
performance requirements, 43, 57
performance testing, 114
performing, 49, 138
phases of, 268, 298
photography, 266
physical, 29, 33, 65, 93, 99, 104, 163, 165, 167, 169–170, 173–177, 179, 187, 189, 213, 248, 255
physical effort, 187
planning, 30, 32, 36, 39–40, 82, 87, 89–90, 93–96, 98, 101, 103–104, 118, 142–144, 147, 183–184, 193, 196, 200, 206, 241, 266, 281
plastic parts, 55
plastics, 43, 55–56, 58, 145, 201
polymers, 58
presentations, 253–254, 256, 259, 261–264, 268–269, 272–273
prices, 34, 84
principles, 57, 61–64, 75, 143, 152, 164, 166, 168, 172–174, 180, 182, 184, 187, 190–192, 199–200, 202, 208–209, 211–212, 224, 228, 246, 298
problem definition, 124–125, 260, 279–280
process, 29–33, 35–39, 41, 43–44, 49, 54–65, 68–71, 74–80, 82–84, 86–87, 89, 92–96, 98, 101, 103–105, 108, 111, 114–115, 117–119, 121–129, 131–137, 141–144, 147–149, 151–156, 158–160, 163–166, 168–169, 171–174, 179, 183–186, 189–190, 192, 195–196, 198, 200, 203, 206–208, 210–211, 216, 218, 221, 226, 231, 236, 241, 243–251, 254–255, 259, 264, 266–267, 269, 273–274, 280–282, 286–287, 291
process capability, 143
process engineering, 75–76, 144, 183
process planning, 143
processes, 32, 35, 44, 54, 56, 58–59, 63–64, 68–69, 76, 89, 118, 124, 140, 143–144, 166, 170–171, 174, 180–182, 184, 187, 196, 198–199, 235, 247, 257, 263, 266–269

processes for, 35, 59, 182, 184
producing, 55, 70
product life cycle, 218, 222, 238
production time, 54, 57
products, 49, 56–57, 64, 68, 91–92, 94–95, 98, 124, 140, 143–144, 164, 169–170, 172, 174–175, 179–181, 183–185, 196–197, 199–200, 209, 218, 233–234, 241, 250, 264, 298
profitability, 152
profits, 148–150, 152
properties, 45, 48–50, 56–57, 79, 129, 136, 138–139
properties of, 56
purchase, 33–34, 83–86, 257
purpose of, 29, 91, 260
purposes, 71, 111, 247, 269, 273

Q

QFD, 257–258, 262, 280, 295
Quality, 56, 75, 77–80, 82, 95, 122, 140–144, 160, 181, 183, 207, 212, 217, 222, 227, 232, 237, 242, 268, 280
of quality, 77, 140, 144
quality, 32, 56–57, 59–62, 64, 69, 71, 75–79, 82, 84, 94–95, 98, 111, 118, 125, 129, 132, 140, 142–144, 168, 174, 177, 179, 181, 189, 199, 223, 260, 268–269, 272, 287–288, 290, 292
quality control, 32, 59–60, 82, 168
quality of life, 174, 179, 189, 199, 223

R

rapid, 41–42, 55, 69, 75, 87, 164, 198
rapid prototyping, 41–42, 55, 69
recycling, 43, 179–182, 199, 219, 245
redesign, 30, 34–35, 39, 62, 65, 83, 123–125, 127–135, 144, 146–148, 156–160, 267–268, 275, 281
reliability, 40, 64, 69–71, 84, 91, 93, 95, 98, 103, 111, 140–142, 144, 152, 280
repairs, 48, 141
reporting, 31, 35, 38, 60, 110, 146, 193
requirements, 29–33, 35–37, 40, 42–43, 54, 57, 59–62, 77, 80, 83, 87, 94–98, 100–103, 107–108, 110–113, 116, 119, 123–125, 130, 132, 136, 138, 140, 143, 154–157, 170, 185, 203–204, 234–236, 246, 250, 254, 270, 280–282, 296
return on investment, 146, 282
review, 33–34, 37, 63, 69, 96, 101, 110, 117, 250, 254, 268, 292
risk assessment, 165, 244, 246
robustness, 141
root causes, 65, 68, 124, 133, 156

S

safety, 43, 47, 49, 64, 69, 77, 79, 91, 93, 95, 98–99, 103, 112, 116, 136, 141, 153–154, 163–169, 174, 190–191, 193, 195, 203–207, 209, 229, 243–246, 248–249, 257, 280–282
safety and, 49, 69, 243
safety, 43, 47, 49, 64, 69, 77, 79, 91, 93, 95, 98–99, 103, 112, 116, 136, 141, 153–154, 163–169, 174, 190–191, 193, 195, 203–207, 209, 229, 243–246, 248–249, 257, 280–282
safety factors, 153–154, 191
scheduling, 38, 82, 95, 103
screws, 56
selection, 30, 45, 49–50, 57, 60, 154, 180, 197, 200–201, 239
Service, 43

service conditions, 45
serviceability, 140, 142
shapes, 37, 44, 55, 91
shortcomings, 123, 225
sketches, 92, 157, 159, 214, 224, 229, 256
slides, 256, 260–261
social issues, 186
social needs, 188
Software, 33, 35–38, 68, 73–74, 80, 102, 107–109, 135, 139, 145–146, 167, 180, 266, 290, 298
software, 29, 33, 35–37, 39, 71–73, 78, 85, 91, 99, 102, 104, 107–108, 135, 138–139, 143, 145, 148, 155, 169, 175, 187, 245, 252, 262, 266–268, 286, 290, 296, 298
software for, 37, 71, 102
software, 29, 33, 35–37, 39, 71–73, 78, 85, 91, 99, 102, 104, 107–108, 135, 138–139, 143, 145, 148, 155, 169, 175, 187, 245, 252, 262, 266–268, 286, 290, 296, 298
soldering, 38
SolidWorks, 80, 140
solving, 76, 90, 137, 144, 202, 253, 275, 296
sources, 181, 200, 263, 270, 272, 282, 291
spare, 142
specifications, 29–31, 33, 36, 39, 54, 56, 77, 79–80, 84, 88, 94–95, 111, 118–119, 123–125, 127–132, 134–135, 140–144, 152, 155–156, 158, 168, 189, 194, 256, 262, 273–275, 280–282
sponsors, 31–33, 35, 38, 40, 42, 63, 77, 123–124, 127, 130, 133–135, 141, 148, 253, 255–256, 258, 261–262, 268, 270, 273, 275, 286–287, 291

stages of, 143, 254, 256, 264
standard components, 54
standardizing, 63
Standards, 94, 104, 168, 203, 208, 213, 218, 228
standards, 32, 40, 77, 79, 93–95, 98–100, 105, 109, 111–112, 116, 118, 125, 132, 136, 139, 141, 143–144, 168, 195–197, 203, 206, 208, 213–214, 216, 218–219, 221–222, 228–229, 231, 234, 257, 274–275, 281
standards for, 141
steel, 45, 57–58
stereolithography, 69, 72
structures, 57, 104, 164, 171, 182, 186–187, 189
subassemblies, 88
subsystems, 30–31, 41, 88–89, 95, 99
suppliers, 33–34, 82, 85, 88, 234, 239, 241, 257
surveys, 281
Sustainability, 43, 128, 188, 191, 198–202, 238–242
sustainable development,, 199
System, 36, 64, 85, 97–98, 102–103, 107–108, 135, 171, 298
systems, 36–37, 39–40, 93–96, 99–100, 129–130, 139, 144, 163–164, 166–169, 171–185, 188, 192–193, 196–202, 205, 247, 258, 267

T

teams, 39–40, 43, 49, 57, 64, 69, 75–79, 83, 86, 125, 129, 132, 136, 144, 148, 153, 155, 172, 255, 259, 261–262, 266, 269
technique, 70, 137, 146
Technology, 55, 69, 176–177, 188, 196–197, 284, 296–297

temperatures, 43, 46, 50, 74–75, 94, 108, 112, 115, 119, 154, 166
tensile strength, 48, 50
test plans, 89, 122
testing for, 40
theories, 190
thermal conductivity, 47
thinking, 195–196, 202, 253, 272, 296, 298
through, 41, 56, 62, 68, 70, 76, 127, 132, 138, 141, 147, 172, 177–178, 180, 183, 185, 188, 195, 199, 219–220, 223, 234, 241, 247, 250, 253, 259, 267–268, 273
time requirements, 157
Tolerances, 56
Tolerancing, 62
tools for, 33, 37, 140
toughness, 47, 49–50
trends, 113, 117–118, 120–122, 174, 178, 201, 270
triple bottom line, 238
types, 43, 62, 64, 74, 79, 83, 137, 155, 185
types of, 43, 62, 83, 185

U

unintended uses, 203
updates, 31, 36, 60, 85, 90, 103, 110, 131, 135, 253–254, 256
user manuals, 32, 167
utility, 191
utility, 191

V

value, 84, 136–137, 141–142, 144, 148–151, 180, 191, 264
values, 45, 96, 100, 121, 137, 150, 186, 192–195
variables, 136–138, 146, 153
Velcro, 81
versus, 38, 43, 49–53
visual appeal, 273

W

warning labels, 167, 205
of wear, 44
wear resistance, 46–47, 56, 112, 115
website, 286
welding, 33, 35, 57–59
welding, 33, 35, 57–59
whistleblowing, 193
Wikipedia, 295

Index

Made in United States
Cleveland, OH
13 January 2025